Circular Dichroism and Magnetic Circular Dichroism Spectroscopy for Organic Chemists

Circular Dichroism and Magnetic Circular Dichroism Spectroscopy for Organic Chemists

Nagao Kobayashi
Department of Chemistry, Graduate School of Science, Tohoku University, Sendai, Japan

Atsuya Muranaka
Institute of Physical and Chemical Research (Riken), Wako-shi, Saitama, Japan

John Mack
Department of Chemistry, Graduate School of Science, Tohoku University, Sendai, Japan

RSC Publishing

ISBN: 978-1-84755-869-5

A catalogue record for this book is available from the British Library

Published by The Royal Society of Chemistry,
Thomas Graham House, Science Park, Milton Road,
Cambridge CB4 0WF, UK

Registered Charity Number 207890

For further information see our web site at www.rsc.org

Printed and bound in Great Britain by CPI Group (UK) Ltd, Croydon, CR0 4YY

Preface

Circular dichroism (CD) and magnetic circular dichroism (MCD) spectroscopy can often provide key information about the conformations and electronic states of chromophore-containing molecules, which is required for a full understanding of the electronic structures and optical spectroscopy. Unfortunately, the theory that underpins these techniques has been largely inaccessible to many organic chemists and biochemists and only a few researchers have carried out detailed quantitative analyses of their spectral data. Until recently, a key problem that has been encountered is that relatively few molecules have been available that can be used to describe the various methods for analysing the spectral data in a clear and concise manner. This is not surprising because people who excel at spectroscopic theory usually lack the skills required to design and synthesise the molecules that would be most appropriate for describing the theory of CD and MCD spectroscopy. Most of the books that have been written on this subject have, therefore, been based on dense sets of mathematical equations and have been aimed primarily at physical chemists and physicists.

Our aim in writing this book is to try to rectify this situation by summarising the different types of CD and MCD spectra and by describing in detail the qualitative and quantitative methods that can be used to analyse the spectral data. Over the last two decades we have successfully synthesised a series of molecules which are ideal for illustrating key points related to the theory of CD and MCD spectroscopy, and we are confident that the time is finally ripe to write a book which will provide the key practical knowledge required to use the CD and MCD techniques to their full potential. In the first chapter, we will provide an introduction to the most important aspects of the theory of electronic absorption, CD and MCD spectroscopy. The important aspects of electronic absorption spectroscopy are described first, since absorption spectra are usually recorded and analysed during any study of the CD and MCD spectra because the spectral bands in each case arise from the same set of transitions. The content is aimed primarily at a reader who already possesses a

Circular Dichroism and Magnetic Circular Dichroism Spectroscopy for Organic Chemists
By Nagao Kobayashi, Atsuya Muranaka and John Mack

Published by the Royal Society of Chemistry, www.rsc.org

first year undergraduate level of understanding of physical chemistry. In the case of small chromophores, the lowest energy electric dipole allowed transitions are often aligned along the direction of the permament dipole moment of the molecule. Transitions of this type are often ideally suited for analysis by CD spectroscopy, since both the electric and magnetic dipole transition moments are well defined in terms of their alignment and can play a key role in generating CD intensity based on several different intensity mechanisms, which will be described in detail. MCD spectra are obtained by placing a magnet in the sample compartment of a CD spectrometer. MCD spectroscopy provides key information about the degeneracy of ground and excited states which cannot easily be derived from the electronic absorption spectrum alone.

In Chapters 2 and 3, we present key information about two of the most widely used analytical approaches that have been reported to date, which could probably be viewed as core knowledge for organic chemists and biochemists who want to become active in this field. Our hope is that in many instances researchers who are new to the field will find that the examples provided can serve as a useful template for analysing their own data. In Chapter 2, the most well known empirical rules for analysing CD spectra, which have been developed over the past 50 years, are described. Although theoreticians have described detailed mathematical rationales for these, there is usually no need to study their research in detail to be able to derive key information about the conformation or configuration of certain specific types of chiral system, such as ketones and substituted benzene rings. The key is to learn how the empirical rules have been applied successfully by earlier researchers and to be fully aware of any exceptions to the rules which have been reported. In Chapter 3, we describe a wide range of examples of the use of exciton coupling theory in the conformational analysis of natural and synthetic dimers, oligomers and polymers. The sign sequence observed in the CD spectra of excition couplets has been found to be related to the relative alignments of the interacting chromophores in space.

In Chapters 4 to 8 more specialised types of analysis are examined. In chapter 4, the use of cyclodextrin inclusion compounds to study the CD spectra of guest molecules trapped in the central cavity is described in detail. Host-guest complexes with clearly defined geometries are readily formed in solution, which enables the analysis of the alignment of band polarisations at wavelengths as short as 200 nm even in the absence of single crystals. The focus in Chapter 5 shifts over to the CD spectroscopy of inorganic complexes based either on the π-system transitions of the ligand or on the $d \rightarrow d$ transitions of the central metal. In recent years the incorporation of 1,1'-binaphthyl (BINAP) moieties has been used to study the asymmetric synthesis of porphyrinoids, since BINAP provides a well defined asymmetric field, which interacts with the porphyrinoid π-system in a manner that can be readily predicted. This research is described in detail in Chapter 6. Although CD spectra can often be readily analysed in conceptual terms based on the theory which describes various intensity mechanisms, in some cases this is not possible and the analysis of the system has to depend almost entirely on theoretical calculations. Molecules with intrinsic chirality are best suited to this approach. Several examples are described in detail in Chapter 7. The results of these calculations have to be treated with caution, however, for reasons that are described in detail. In Chapter 8, the CD spectroscopy of biomolecules will be described. Since large proteins contain many chromophores, which lie at various distances from each other in a wide range of alignments, the quantitative analysis of these spectra is usually not possible, but the CD technique can still be used to derive qualitative information and has been widely applied on this basis.

In Chapter 9, representative examples of the use of MCD spectroscopy are described. Many of the examples revolve around the study of porphyrin complexes, since the technique has been particularly widely utilised in this context. The theory of MCD spectroscopy can be challenging for many organic chemists, because it revolves around the application of molecular orbital theory to conceptualise the structure and bonding of chromophores, rather than the valence bond theory approach that tends to be used in organic chemistry. In Chapter 10, we describe how Michl's perimeter model, a molecular bonding theory approach, can be used a molecular bonding theory approach can be used to analyse the MCD spectra of aromatic and antiaromatic π-systems.

Contents

Circular Dichroism and Magnetic Circular Dichroism Spectroscopy for Organic Chemists
By Nagao Kobayashi, Atsuya Muranaka and John Mack

Published by the Royal Society of Chemistry, www.rsc.org

Abbreviations

[3,3]PCP	[3,3]paracyclophane
acac	acetylacetonate
Ara	arabinose
BINAP	1,1'-binaphthyl
bipy	2,2'-bipyridyl
B3LYP	Becke 3-parameter (exchange), Lee, Yang and Parr (correlation)
CAM–B3LYP	Coulomb attenuated method B3LYP
CBD	cyclobutadiene
CD	circular dichroism
CI	configuration interaction
con.	consignate
Cp	cyclopentadiene
CyD	cyclodextrin
dis.	dissignate
DFT	density functional theory
DNA	deoxyribonucleic acid
edtm	electric dipole transition moment
en	ethylenediamine
EPR	electron paramagnetic resonance
Fc	ferrocene
Gal	galactose
Glu	glucose
HPLC	high-performance liquid chromatography
ICD	induced circular dichroism
LCAO	linear combination of atomic orbitals
lcp	left circularly polarised
LMCT	ligand to metal charge transfer
Lys	lysine

Circular Dichroism and Magnetic Circular Dichroism Spectroscopy for Organic Chemists
By Nagao Kobayashi, Atsuya Muranaka and John Mack

Published by the Royal Society of Chemistry, www.rsc.org

Man	mannose
MCD	magnetic circular dichroism
mdtm	magnetic dipole transition moment
MO	molecular orbital
MORD	magnetic optical rotatory dispersion
MLCT	metal to ligand charge transfer
MW	molecular weight
NMR	nuclear magnetic resonance
OAM	orbital angular momentum
OBz	benzoate
ORD	optical rotatory dispersion
Pc	phthalocyanine
phen	*o*-phenanthroline
rcp	right circularly polarised
RNA	ribonucleic acid
SCF-CI-DV	self consistent field–configuration interaction–dipole velocity
SCF-PPP-CI	self-consistent field Pariser-Parr-Pople configuration interaction
SM	spectroscopic moment
TD-DFT	time dependent–density functional theory
TFA	trifluoroacetic acid
TZVP	triple zeta valence plus polarisation
VTVF	variable temperature variable field
x–c	exchange–correlation
Xyl	xylose
ZFS	zero field splitting

CHAPTER 1

Theory of Optical Spectroscopy

1.1 Electronic Absorption Spectroscopy

Optical spectroscopy is based ultimately on the interaction between atoms or molecules and incident electromagnetic radiation. In the 1860s, a Scottish physicist called James Clerk Maxwell first postulated that an oscillating electric field generates an oscillating magnetic field and *vice versa*. A propagating sinusoidal electromagnetic wave can be formed on this basis, with electric and magnetic fields oscillating perpendicular to one another and to the direction of propagation. Electromagnetic radiation exhibits both wave properties and particle properties and was described successfully by Albert Einstein in quantum mechanical terms as a particle, referred to as a photon (hν), which has no mass or charge. At longer wavelengths in the IR region ($>$1000 nm) the interaction between the atomic nuclei and the oscillating electric and magnetic fields typically results in molecular vibrations, which can be studied by infrared spectroscopy. At shorter wavelengths the heavier nuclei can no longer oscillate significantly, but the surrounding cloud of electron density can still be polarised in the direction of the oscillating electron field, resulting in an electronic transition from a groundstate electron configuration to an excited state. UV-visible absorption spectroscopy can be used to derive key information about the electronic structures of molecules on this basis, while techniques such as fluorescence spectroscopy can be used to derive information from the manner in which the molecule returns to its groundstate configuration.

As shown in Figure 1.1, molecular orbital theory can be used to describe this electronic excitation on the basis of the transfer of an electron from an occupied molecular orbital to an unoccupied molecular orbital. The energy difference between the ground and excited states (ΔE) is proportional to the frequency of the absorbed electromagnetic radiation (ν):

$$\Delta E = \mathrm{h}\nu = \mathrm{hc}/\lambda \tag{1.1}$$

Circular Dichroism and Magnetic Circular Dichroism Spectroscopy for Organic Chemists
By Nagao Kobayashi, Atsuya Muranaka and John Mack

Published by the Royal Society of Chemistry, www.rsc.org

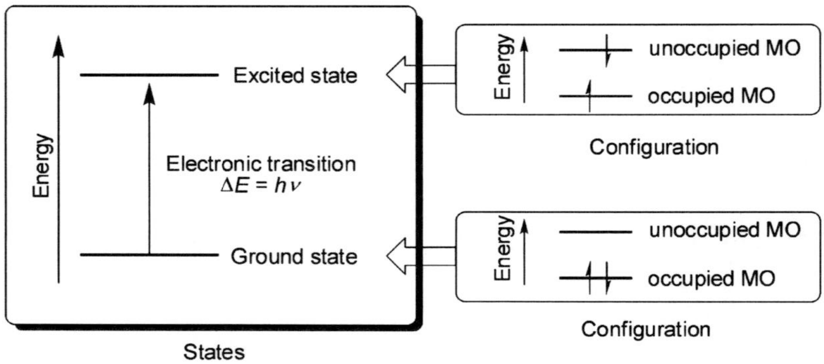

Figure 1.1 Energy diagram showing an electronic transition.

where h is the Planck constant ($h = 6.626 \times 10^{34}$ J s), and c and λ denote the velocity of light ($c = 2.998 \times 10^{8}$ m s^{-1}) and the wavelength, respectively. Especially in the context of organic molecules, absorption in a particular region of the spectrum is often characteristic of a transition that is associated primarily with a particular type of bond or structural unit within a molecule. These structural units are usually referred to as chromophores. In the context of saturated organic molecules, wavelengths much shorter than 200 nm are required to cause electronic transitions. Since the conventional use of UV-visible absorption spectrometers under an air atmosphere tends to be limited to 200 nm, owing to strong absorption by oxygen and ozone formation at shorter wavelengths, electronic absorption spectroscopy is applied primarily to the π-systems of organic molecules and inorganic metal complexes, which absorb strongly at wavelengths >200 nm, with a particularly strong focus on aromatic and heteroaromatic cyclic compounds.

The interaction of UV or visible region light (typically 200–750 nm) with a molecule or complex can result in an electronic excitation from one molecular orbital to another, resulting in a transition from the groundstate electronic configuration to an excited state, Figure 1.1. This inherently results in a rearrangement of the electron density of a molecule. Since the size of molecules and complexes will typically be a few orders of magnitude smaller than the wavelength of UV-visible light, the electric field induces an oscillating electric dipole moment upon absorption of a photon. An electric dipole transition moment (edtm), $\boldsymbol{\mu}$, can be defined for each transition, which describes the net linear displacement of charge during a transition. The initial point of this vector is set to the centre of gravity of the molecule, and the square of the transition dipole moment is proportional to the intensity of the electronic transition. If a transition is dipole forbidden, $\boldsymbol{\mu} = 0$, while a transition is said to be allowed if $\boldsymbol{\mu} > 0$. The direction in which $\boldsymbol{\mu}$ is aligned determines the polarisation of the associated spectral band with respect to the *x*-, *y*- and *z*-axes. It should be noted that edtms are different from static electric dipole moments (also known as permanent dipole moments), which describe the polarisation of charge in a

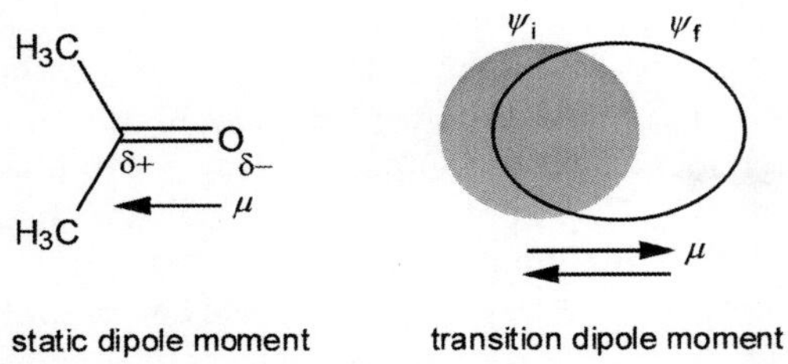

Figure 1.2 An illustration of a static dipole moment (left) and an electric transition dipole moment (right). The grey and white circles indicate electron distributions of the groundstate and an excited state.

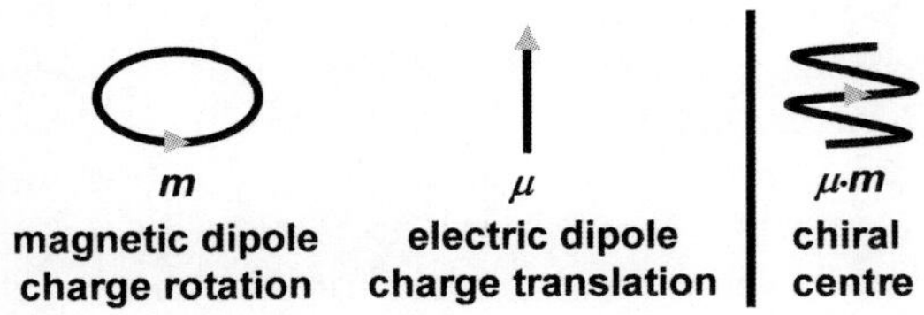

Figure 1.3 The origin of chirality, based on the combined effect of electronic and magnetic transition moments.

molecule in the groundstate, Figure 1.2. The direction of a static dipole moment can be determined definitively on the basis of the molecular structure. In contrast, since a transition dipole has an oscillating property, the choice of the sense of a transition dipole moment is arbitrary and depends on the phase of the wavefunctions. A magnetic dipole transition moment (mdtm), ***m***, can also be defined for each electronic transition, which describes the net circulation of charge during a transition, Figure 1.3. The edtms are usually the dominant factor in coupling the groundstate with excited states within UV-visible absorption spectroscopy, since they tend to be *ca.* five orders of magnitude stronger than magnetic dipole moments. It should be noted that this is not the case during the quantitative analysis of CD spectral data, since the intensity mechanism is based on an interaction between electric and magnetic dipole transition moments. Sections 2.2 and 5.2 describe the analyses of carbonyl $n \rightarrow \pi^*$ transitions in organic molecules and the $d \rightarrow d$ transitions of transition metal complexes, which are magnetic dipole allowed but electric dipole forbidden.

Group theory can be used to determine whether transitions are electric dipole and magnetic dipole allowed or forbidden and the polarisation of the spectral bands, which arise in the UV-visible absorption spectrum based on the value of the transition moment integral:

$$\int \psi_1 \boldsymbol{\mu} \psi_2 \mathrm{d}\tau \qquad (1.2)$$

where ψ_1 and ψ_2 are the wavefunctions of the ground and excited states, respectively, and $\boldsymbol{\mu}$ is the transition moment operator, which transforms in different ways along the *x*-, *y*- and *z*-axes depending on the point group. The symmetry characteristics of each component can be obtained from standard character tables based on the point group to which the molecule or complex belongs. If the value of this integral is zero, the transition is forbidden. The integral itself does not need to be calculated, since character tables can be used to determine the symmetry of the transition moment function, $\psi_1\, \boldsymbol{\mu}\, \psi_2$. If the symmetry of this function spans the totally symmetric representation of the relevant point group, the value of the integral is nonzero and the transition is electronically allowed. If not, although the transition is electronically forbidden it may still gain significant intensity under certain circumstances based on vibronic interactions with close-lying allowed transitions.

In UV-visible absorption spectra, either absorbance (A) or molar absorptivity (ε) are usually used as the *y*-axis unit for intensity, while either a wavelength (λ) or a wavenumber (cm^{-1}) scale is used to plot the *x*-axis. A wavelength scale tends to be the norm in the literature. It is important to bear in mind that this is inversely proportional to an energy scale, however. Some authors prefer to use a wavenumber scale for the *x*-axis, so that the relative energies of the spectral bands can be visualised readily. Absorbance (A) is defined to be the logarithm of the ratio of the incident (I_0) and transmitted (I) radiation.

$$A = \log_{10}(I_0/I) \tag{1.3}$$

The relationship between A and ε ($mol\ l^{-1}\ cm^{-1}$) is described by Beer-Lambert's law:

$$A = \varepsilon c l \tag{1.4}$$

where c is the concentration of the molecule, expressed in mol. l^{-1} and l is the pathlength in cm.

A key question that needs to be addressed in any detailed analysis of not just the UV-visible absorption spectrum, but also the CD and MCD properties, is the extent to which a particular transition will absorb incident radiation, and why. In some instances molar absorptivities can be very large, for example those of the electric dipole allowed $\pi \rightarrow \pi^*$ bands associated with the π-conjugation systems of aromatic compounds (>10 000), but in others, such as bands arising from the $d \rightarrow d$ transitions of metal complexes, which are electric dipole forbidden (*ca.* 100), they are very small. The overlap of the orbitals involved in the electronic excitation is often a key factor in this regard. A typical absorption spectrum is shown in Figure 1.4. The absorption band with a λ_{max} value of about 300 nm is associated with the HOMO (π) $\rightarrow$ LUMO (π^*) transition and is polarised parallel to the long axis of the molecule. Since the energy gap between the HOMO and LUMO decreases with increasing π-delocalisation, the electronic spectra of organic molecules often provide an

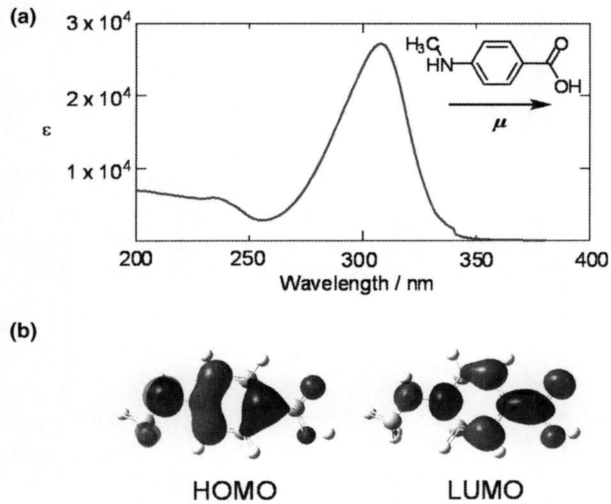

Figure 1.4 (a) Electronic absorption spectrum of 4-(methylamino)benzoic acid measured in acetonitrile. The arrow indicates the alignment of the edtm of the lowest-energy transition. (b) The frontier MOs plotted at an isosurface value of 0.02 a.u. derived from a geometry optimisation using the B3LYP functional with 6–31G* basis sets.[28]

indication of the size of an aromatic or heteroaromatic π-system and its degree of delocalisation. Conjugated systems such as anthocyanins, carotenoids and chlorophylls have relatively small HOMO–LUMO band gaps and are strongly coloured because they absorb in the visible region.

With the recent development of commercially available software packages time-dependent density functional theory (TD-DFT) calculations can now be carried out routinely, so experimental spectral data can usually be compared readily to a calculated spectrum and, at least in the case of smaller molecules, bands can often be assigned to specific electronic transitions on this basis. For larger complexes and molecules, TD-DFT calculations typically only provide a reliable description of absorption bands in the lower energy region of electronic absorption spectra, because there tends to be extensive configuration interaction between the electronic states in the higher energy region. This means that the main focus of quantitative analyses often tends to be the CD and MCD signals associated with the lower energy transitions. It is not possible to calculate ε_{max} values based on TD-DFT, because the spectral bands associated with each transition vary markedly in width in a manner that cannot be readily predicted, owing to the excitation of molecular vibrations and the effect of Heisenberg uncertainty on the lifetime of the excited states. Theoretical approaches, therefore, tend to focus on the integrated intensity of the entire spectral band rather than the ε value at a discrete wavelength. When the absorption spectrum is plotted using wavenumber units, the total absorption under a band can be expressed as an integral based on an energy rather than a wavelength scale:

$$f = 4.32 \times 10^{9} \int \varepsilon \mathrm{d}\nu \qquad (1.5)$$

Integration (dν) over the entire spectral band results in an oscillator strength value, *f*, which provides a measure of the probability that a molecule will absorb incident electromagnetic radiation over a certain energy range. The oscillator strength can be approximated by viewing the spectral band as a triangle:

$$f \approx 4.32 \times 10^{9} \varepsilon_{\mathrm{max}} \Delta\nu_{1/2} \qquad (1.6)$$

by multiplying the ε_{max} value at the band centre by the width of the band at half height, $\Delta\nu_{1/2}$. Calculations of the UV-visible absorption spectra based on density functional theory (DFT) typically provide oscillator strength values, *f*, for each spectral band, since the oscillator strength values are related to the transition dipole moments and hence can be calculated from the wavefunction of the molecule.

A related unit referred to as the dipole strength, *D*, based on the square of the edtm,[1] is sometimes used instead:

$$D = \langle \Psi_0 | \boldsymbol{\mu} | \Psi_a \rangle^2 = \mathrm{F} \int \varepsilon \mathrm{d}\nu \qquad (1.7)$$

where F is a factor that depends on the units used and conventions followed. In the context of MCD spectroscopy, we recommend the use of a definition proposed by Piepho and Schatz,[2] which is denoted as $\mathcal{D}_0$, for reasons that will be described in Section 1.3.3:

$$\mathcal{D}_0 = 3.062 \times 10^{-3} / \nu_{\mathrm{max}} \int \varepsilon \mathrm{d}\nu \approx \langle A \rangle_0 / 326.6cl \approx \langle \varepsilon \rangle_0 / 326.6cl \qquad (1.8)$$

where $\mathcal{D}_0$ is expressed in (Debye)^2 units (D^2), ν_{max} is the band centre energy in cm^{-1} and $\langle A \rangle_0$ is the zeroth moment of the intensity for the entire absorption band. The equation on the right-hand side is valid for any symmetrical band shape.

1.1.1 Exciton Coupling Theory

The molecular exciton model describes the coherent superposition of the excited states of aggregated systems. When there are two or more close-lying and strongly absorbing chromophores, the edtms can couple, leading to a splitting of the excited state energies to an extent that markedly modifies the spectral properties. The exciton concept is often used to analyse the electronic absorption and CD spectra of aggregated chromophore systems, which are not necessarily chemically bonded. Since excitonic interactions usually provide the most significant contributions to the CD spectra of these systems, exciton coupling theory is particularly important for analysing their CD spectra. Details of the theory of exciton coupled CD and its practical application are discussed in Section 1.2.2 and in Chapter 3. This is in marked contrast to MCD spectroscopy where it can usually be disregarded. For example, Hughes *et al.*[3] recently demonstrated that the exciton coupling contribution to the MCD signal of chlorophyll dimers has a second order effect, since it only influences the observed MCD intensity by a few percent.

The molecular exciton model is a state interaction perturbation theory that deals with the excited state resonance interaction between chromophores in weakly coupled electronic systems. The theory was originally developed by Davydov to explain the absorption spectra of molecular crystals (van der Waals crystals).[4] Kasha[5] successfully applied the exciton model subsequently to molecular spectroscopy based on a vector model approach, which can be used to depict how the exciton interaction energy term, and the edtm for the exciton states of dimers and higher oligomers, may be deduced from vector diagrams of the interaction between the transition moments of the component chromophores. Although the exciton model can be described rigorously on the basis of quantum mechanics, such a description is not required for chemists and biochemists to be able to analyse their data successfully. The application of molecular exciton theory requires spatially separated chromophore units. In a chromophore dimer system in which each chromophore has an edtm ($\boldsymbol{\mu}_n$), the exciton coupling energy between the two chromophores (V_{12}) is based on a coulombic interaction energy and can be described by the following equation using a point-dipole approximation:

$$V_{12} = \frac{1}{4\pi\epsilon_0}\left\{\frac{\boldsymbol{\mu}_1 \cdot \boldsymbol{\mu}_2}{(\mathrm{R}_{12})^3} - 3\frac{(\boldsymbol{\mu}_1 \cdot \mathrm{R}_{12})(\boldsymbol{\mu}_2 \cdot \mathrm{R}_{12})}{(\mathrm{R}_{12})^5}\right\} \tag{1.9}$$

$$= \frac{1}{4\pi\epsilon_0}\frac{\boldsymbol{\mu}^2}{(\mathrm{R}_{12})^3}(\sin\alpha\sin\gamma\cos\tau + 2\cos\alpha\cos\gamma) \tag{1.10}$$

where ϵ_0 is a physical constant for vacuum permittivity, while R_{12} is the vector from chromophore 1 to chromophore 2 and α, γ and τ are angles between the vectors defined in Figure 1.5. It should be noted that the angle τ is always taken in an anticlockwise direction when the system is viewed from chromophore 1 to chromophore 2. The exciton energy for a degenerate system should be inversely proportional to the cube of the interchromophoric distance (R_{12}) and proportional to the square of the transition moment ($\boldsymbol{\mu}$) of the interacting chromophore. The point-dipole approximation holds when the separation of the chromophores is much larger than the transition dipole lengths, Figure 1.6. When the interchromophoric separation is small, a higher level of approximation, such as the monopole approximation, is required to describe the exciton interaction energy.

As shown in Figure 1.7, in the context of a dimer the in-phase and out-of-phase arrangements of the transition dipoles lead to two exciton states.

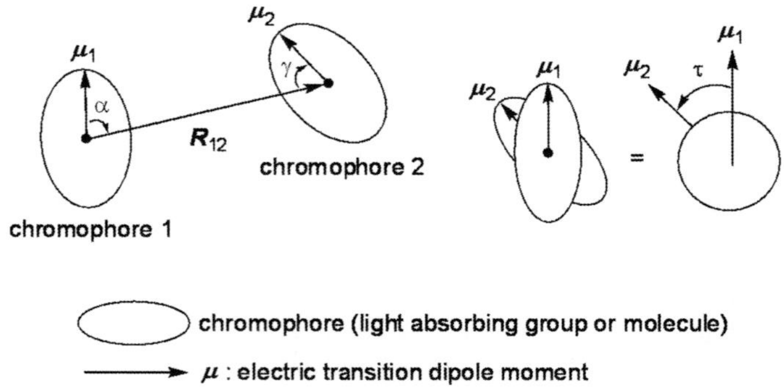

Figure 1.5 Geometrical parameters for a chromophoric dimer system. The ovals correspond to a light absorbing unit or molecule. It should be noted that the angle τ is always taken in an anticlockwise direction when the system is viewed from chromophore 1 to chromophore 2.

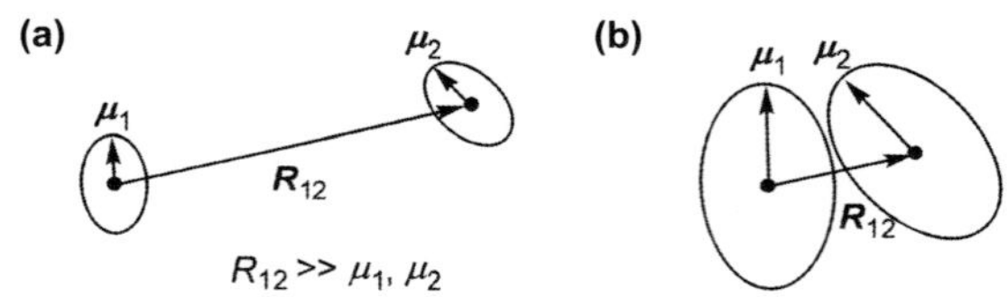

Figure 1.6 Chromophoric dimer with large and small interchromophoric distances. The point-dipole approximation is applicable in the case of (a).

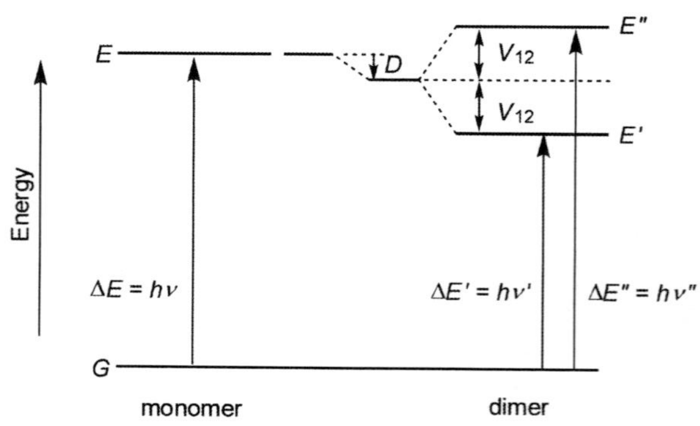

Figure 1.7 Exciton resonance splitting in a chromophoric dimer. *D* indicates the van der Waals energy-lowering term.

The energy splitting caused by the excitonic interaction ($2V_{12}$) is referred to as a Davydov or exciton splitting. The mutual orientation of the chromophores plays an important role in determining the magnitude of the energy splitting. Kasha's exciton model[5] enables a simple vector model analysis of the exciton band structure of spectral changes in molecular dimers, oligomers and polymers. The relative changes in the energy levels of aggregated molecular systems can usually be deduced from the electrostatic interactions between the edtms. The simplest systems to analyse are degenerate chromophoric dimers. When two chromophore units are located in a face-to-face arrangement, Figure 1.8, electrostatic attraction and repulsion between the two dipoles gives rise to two exciton states E' and E'' symmetry. The net transition dipole moment ($\boldsymbol{\mu}_{net}$), based on the vector sum of the two component transition moments, is zero for the lower exciton state (E'), so the electronic transition from the

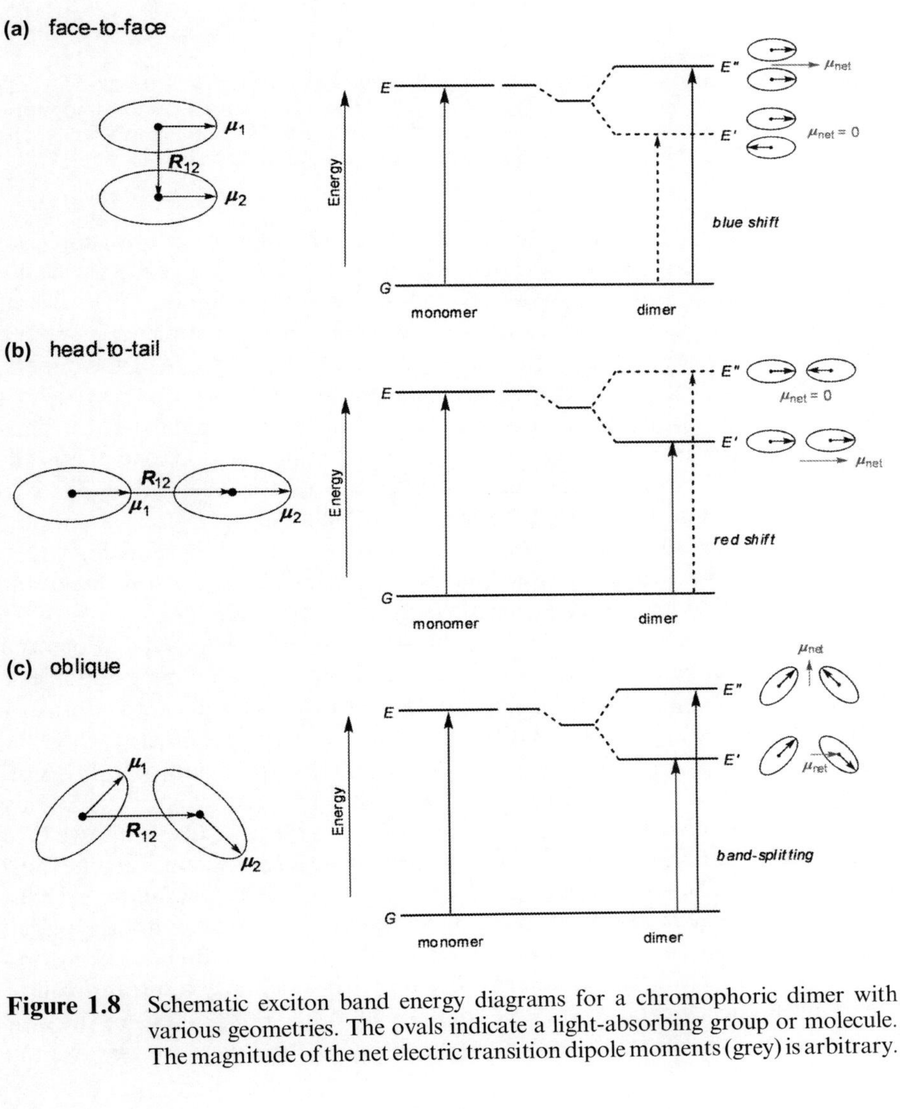

Figure 1.8 Schematic exciton band energy diagrams for a chromophoric dimer with various geometries. The ovals indicate a light-absorbing group or molecule. The magnitude of the net electric transition dipole moments (grey) is arbitrary.

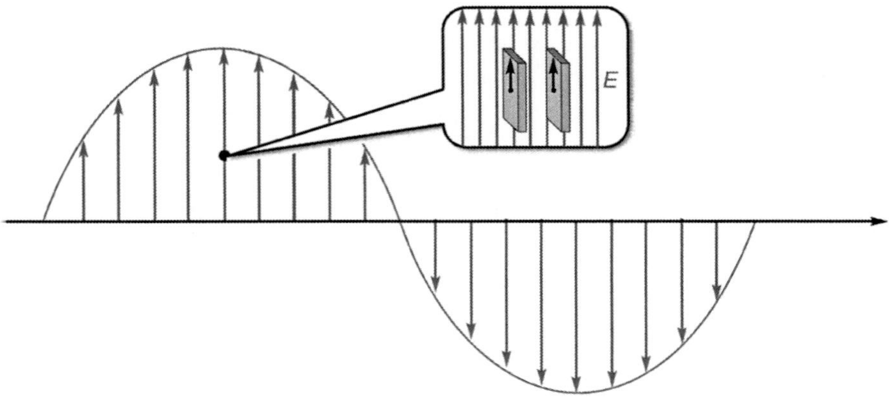

Figure 1.9 A diagrammatic illustration of the relative magnitudes of the electric field of the incident electromagnetic radiation and the edtms in a chromophoric dimer. Note that both chromophoric units are simultaneously under the influence of what is effectively an identical electric field.

groundstate to the E' state is dipole forbidden. The higher exciton state (E'') has a non-vanishing $\boldsymbol{\mu}_{net}$, so the transition to the E'' state is electronically allowed. The origin of these selection rules can be rationalised on the basis of consideration of the relative size of the electric field and the dimer, Figure 1.9. The excitation wavelength is generally much larger than the size of the dimers, so an identical electric field can be assumed during electronic excitations. As a result, the absorption band of a face-to-face dimer is predicted to be blue shifted to higher energy relative to that of the monomer. In the case of a head-to-tail dimer (Figure 1.8), however, a red shift of the absorption band to lower energy is predicted, since the lower and higher exciton states are dipole allowed and forbidden, respectively. If the system has an oblique arrangement, Figure 1.8, both the exciton states become dipole allowed and a splitting of the absorption band into two components is usually observed.

Cyclic trimer and tetramer systems such as cyclopropenyl cation and cyclobutene can be used as examples of what happens with higher oligomeric systems, Figure 1.10. The lowest exciton state of the cyclic trimer is optically forbidden owing to the cancellation of the net transition moment, while the transition to the doubly degenerate higher energy state is dipole allowed. Three exciton states are anticipated for the cyclic tetramer, Figure 1.10. Only the transition to the doubly degenerate E_2 state is electric dipole allowed, so only minor spectral changes are typically observed for this geometrical arrangement. The exciton energy diagrams of polymeric systems, such as linear polymers, can also be derived by applying the exciton coupling model, Figure 1.11. A quasi-continuous exciton band is anticipated. If transition dipoles are aligned perpendicular to the polymer axis, only the transition to the highest-energy exciton state is optically allowed. In contrast, in the case of a head-to-tail polymer, only the transition to the lowest-energy exciton

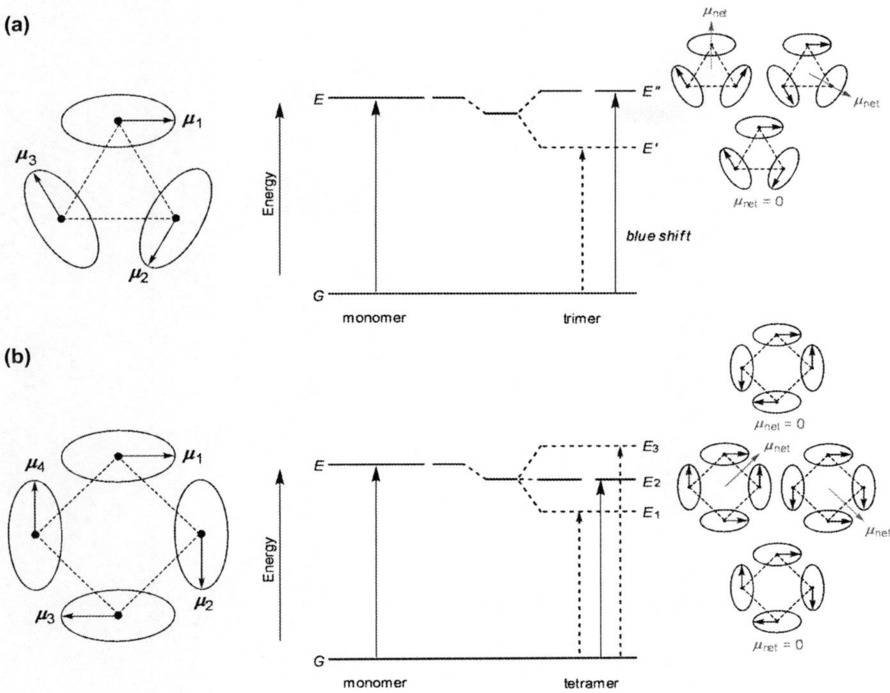

Figure 1.10 Schematic exciton band energy diagrams for cyclic trimer (a) and tetramer (b). The ovals indicate a light-absorbing group or molecule. The magnitudes of the net edtms (grey) are diagrammatic only.

state is dipole allowed. Face-to-face and head-to-tail polymers which display these spectral properties are usually referred to as H- and J-aggregates, respectively.

1.2 CD Spectroscopy

Analysis of optical spectral data rests on the fact that the absorption or emission of a photon of light with specific wavelength properties results in a redistribution of charge within a molecule, based on a transition moment, which couples the ground and excited electronic states, resulting in a new stationary electronic state, see Figure 1.3. The electric field of the electromagnetic radiation can cause a linear displacement of the electronic charge, creating an edtm, while the magnetic field can cause a circulation of charge, creating a mdtm. In the absence of a mirror plane or centre of symmetry, these two motions can combine to cause a helical redistribution of the electron density. Since incident photons of left (lcp) and right (rcp) circularly polarised light are chiral in their handedness properties, Figure 1.12, they interact to differing extents during the electronic excitation of chiral molecules by specific

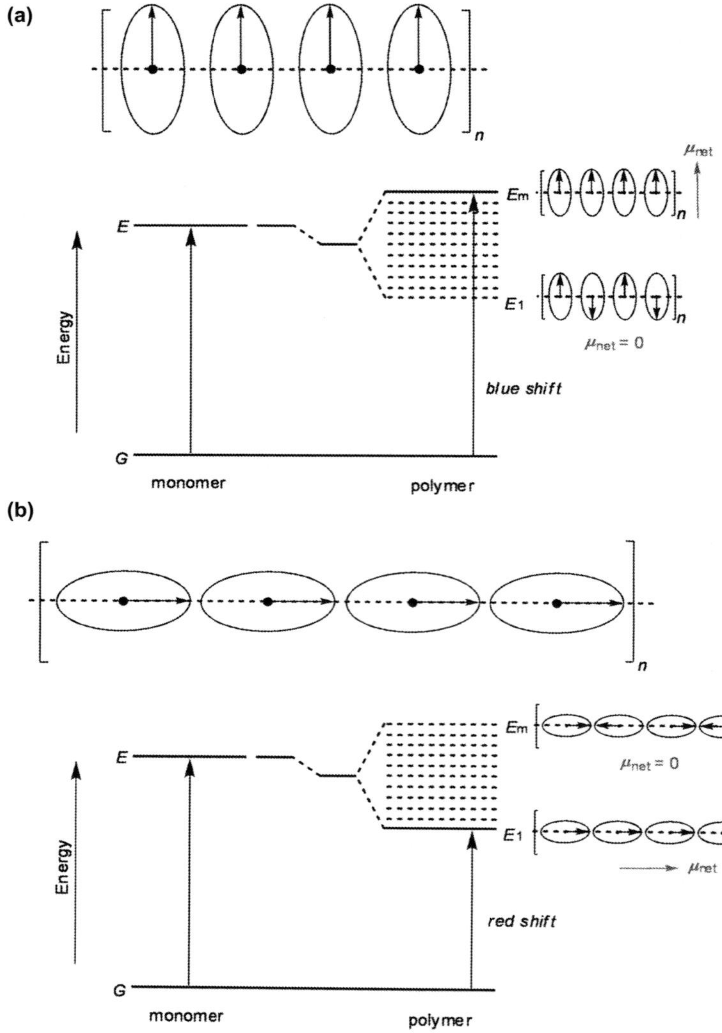

Figure 1.11 Schematic exciton band energy diagrams for linear polymers: (a) face-to-face type polymer; (b) head-to-tail type polymer. The ovals indicate a light-absorbing group or molecule. The magnitudes of the net edtms (grey) are diagrammatic only.

electronic transitions. In the context of organic molecules, a carbon with four different substituent groups is referred to as a chiral centre. Stereoisomers of a chiral molecule will preferentially absorb either lcp or rcp light based on the handedness of the helical electron redistribution for a particular transition. This differential absorbance of lcp and rcp is often referred to as the Cotton effect and forms the basis of CD spectroscopy. Later chapters will describe how

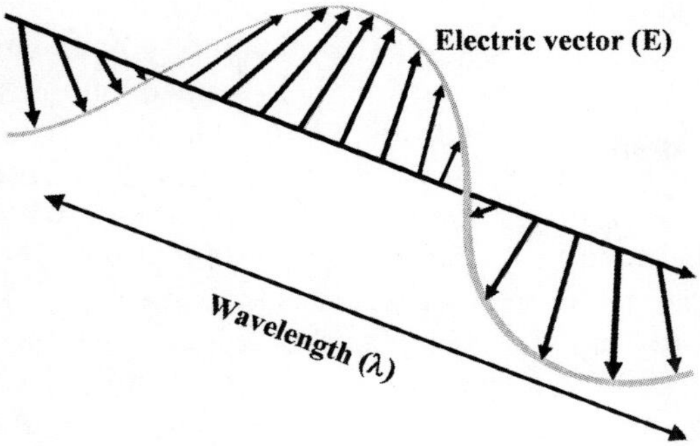

Figure 1.12 The electric vector of lcp light follows an anticlockwise path, rotating once per wavelength along the axis of light propagation.

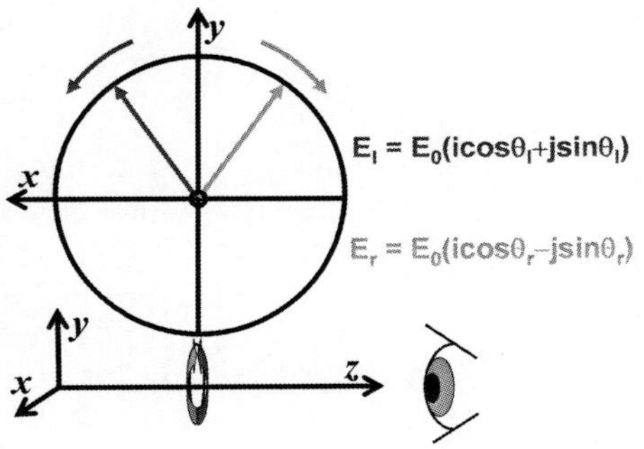

Figure 1.13 The original definition of lcp and rcp within CD spectroscopy.[6] Light plane polarised in the *yz*-plane can be described in mathematical terms as being comprised of lcp or rcp components of equal phase and amplitude (top). The perspective towards the source is also depicted (bottom).

detailed structural information about the relative arrangement of the atoms of chiral molecules in three-dimensional space can be derived by analysing the information about the helical motion of electron density upon electronic excitation, which can be derived from CD spectra.

One of the major problems faced in any description of optical activity is that differing left- and right-handedness conventions have been adopted depending on the context. The literature for CD and MCD spectroscopy can be very confusing unless the reader firmly grasps that the classical optics definition typically used by chemists when analysing CD and MCD spectra is towards the light source with the axis of light propagation being defined by the *z*-axis, Figure 1.13, as opposed to the perspective along the axis of light propagation that is often favoured by physicists. A clockwise right-handed perspective within one definition corresponds to an anticlockwise left-handed definition in the other, and since a systematic set of conventions has never been introduced textbooks on the subject can adopt either definition, often with no explanation of this potential source of confusion. Left-handed photons were originally defined as positive, since the electric vector rotates anticlockwise from the positive y-axis to the positive x-axis within a right-handed set of axes from positive y-values to positive x-values.[6] Unfortunately, the original definition for lcp light is the same as that used to define the right-handedness of the α-helix of DNA (deoxyribonucleic acid),[7] where the perspective is from the source along the axis of propagation. The fact that the left- or right-handedness of a helix is independent of perspective when viewed from either end complicates matters still further in conceptual terms, since it can provide the impression that handedness properties are always completely independent of perspective.

1.2.1 Units used for CD Spectroscopy

In the present day, CD intensity is usually reported as the differential absorbance of left (A_L) and right circularly polarised (A_R) light and can be expressed as:

$$\Delta A = A_L - A_R \tag{1.11}$$

As is the case with absorption spectroscopy, shown in equation (1.4), molar extinction coefficients (Δε) can be reported by taking the cell pathlength and the molar concentration of the compound into account:

$$\Delta\varepsilon = \varepsilon_L - \varepsilon_R = \Delta A / cl \tag{1.12}$$

so that ε_L and ε_R are the molar extinction coefficients for lcp and rcp light, c is the molar concentration and l is the pathlength in centimetres.

Although ΔA and $\Delta\varepsilon$ are the most readily accessible intensity units for chemists and biochemists in conceptual terms, since A and ε are used in UV-visible absorption spectroscopy, for historical reasons CD spectra were initially described in terms of degrees of ellipticity (θ), Figure 1.14, due to the legacy of an older spectroscopic technique called optical rotatary dispersion (ORD) spectroscopy, which is based on the rotation of a beam of plane polarised light. Prior to the emergence of photoelastic modulators in the 1960s, which enable the formation of an alternating beam of lcp and rcp light from plane polarised light, ORD spectroscopy was the primary technique used to study optical activity. Data expressed in units of ellipticity could be readily compared to the existing literature, since both ellipticity and optical rotation provide measures of changes in the polarisation state of a linearly polarised beam of electromagnetic radiation. CD data reported using ellipticity units are still frequently encountered in the literature so this remains more than a historical curiosity.

Plane polarised light can be viewed as a superposition of lcp and rcp light of equal amplitude and phase, since a projection of the combined amplitudes of these two components perpendicular to the axis of light propagation yields a

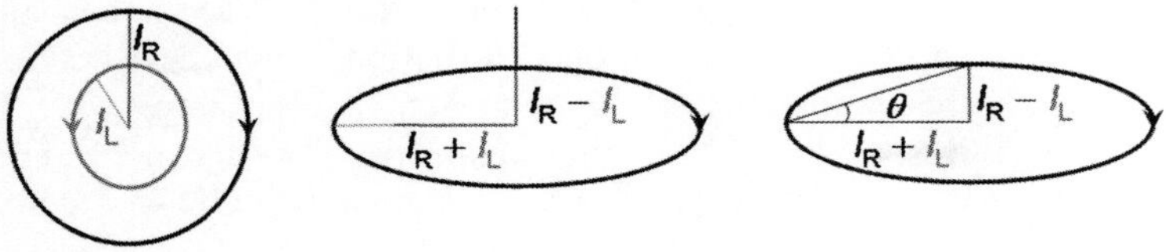

Figure 1.14 When lcp and rcp light have different intensities (I_L and I_R) they combine to form an elliptically rather than linearly polarised beam. The ellipticity, θ, is defined as the angle between the long and short axes of the ellipse.

straight line. When this linear polarised beam passes through an optically active sample, the amplitude of the more strongly absorbed component will be smaller than that of the less strongly absorbed component, resulting in an ellipse when the amplitudes of the components are combined, Figure 1.14. The degree of ellipticity is defined as the tangent of the ratio of the minor to major elliptical axes. Linearly polarised light has 0° of ellipticity, while fully lcp and rcp light have +45° and −45° ellipticity, respectively. The ellipticity introduced by circular dichroism is not directly related to optical rotation of the polarisation plane, which is associated with a difference in the refractive index, n, due to a change in the phase of the lcp and rcp components and is usually referred to as circular birefringence. When CD exists, optical rotation must also exist, however, so CD and ORD spectral data can be directly related to each other by carrying out Kramers-Kronig transformations. Circular dichroism is observed only at wavelengths where a chiral molecule absorbs light, while at all other wavelengths it is optical rotation that is observed.

The interconversion between θ and ΔA is relatively straightforward:

$$\Delta A = \theta / 32.982 \tag{1.13}$$

Owing to the small size of most ellipticity measurements, θ is often reported in millidegrees and a conversion factor of 32 982 is used instead in this context. Molar ellipticity values can be derived in much the same way that $\Delta\varepsilon$ values are derived from data sets expressed in units of ΔA. When the molar concentration and cell pathlength are introduced according to Beer-Lambert's law, units of molar ellipticity can be derived, $[\theta]_M$. Following polarimetric conventions, molar ellipticity is reported in deg.cm^2.dmol^{-1} or deg.M^{-1}.m^{-1} and molar ellipticity is calculated as:

$$[\theta]_M = 100\theta / cl \tag{1.14}$$

and molar extinction coefficient and ellipticity values can be converted by:

$$\Delta\varepsilon = [\theta]_M / 3298.2 \tag{1.15}$$

The conversion factor is 100 times larger than that between absorbance and ellipticity in this context, since the molar extinction convention defines pathlength in centimetres, while that of ellipticity defines it in metres.

Just as the corresponding UV-visible absorption band can be defined in terms of a dipole strength based on $\boldsymbol{\mu}^2$, a spectral band in a CD spectrum can be characterised by its rotational strength, which is given by:

$$R = 3\mathrm{h}c10^3 \ln(10)/32\pi^3\mathrm{N_A} \int \Delta\epsilon(\nu)\mathrm{d}\nu/\nu \tag{1.16}$$

Rotational strengths, like dipole strengths, are the product of wavefunction overlap integrals and can be expressed as the Rosenfeld equation:

$$R = \mathrm{Im}[\langle\Psi_\mathrm{o}|\boldsymbol{\mu}|\Psi_\mathrm{a}\rangle\cdot\langle\Psi_\mathrm{a}|\boldsymbol{m}|\Psi_\mathrm{o}\rangle] \tag{1.17}$$

where Im refers to the imaginary component of the scalar product between the electric and magnetic dipole moments. Although $\boldsymbol{\mu}$ is a real vector, $\boldsymbol{m}$ is purely imaginary, since it describes the rotation of the coordinate system so that $z = x + iy$. In most contexts it can be stated that:

$$R = \boldsymbol{\mu}\cdot\boldsymbol{m}\cos\theta \tag{1.18}$$

where θ is the angle between the two dipoles, Figure 1.15. In recent years, it has become possible to derive values for the rotational strengths of spectral bands from commercially available DFT software packages. If a molecule has a plane or centre of symmetry, the sum of all of the induced electric and magnetic dipoles will either be zero, or the vectors of the magnetic and electric dipoles will be orthogonal to each other resulting in no optical activity, since $\cos\theta = 0$ if $\theta = 90°$. In later chapters, we will examine in detail how CD spectroscopy can be used to carry out detailed analyses of different types of molecular system, where the magnetic and electric dipole transition moments are not orthogonal to each other and strong CD signals are generated.

Since the order of magnitude of the edtm is typically 1 Debye (μ_D), while that of $\boldsymbol{m}$ is approximately 1 Bohr magneton (μ_B), the order of magnitude of R is usually of the order of 0.927×10^{-24} J T^{-1}. In the older literature this would be

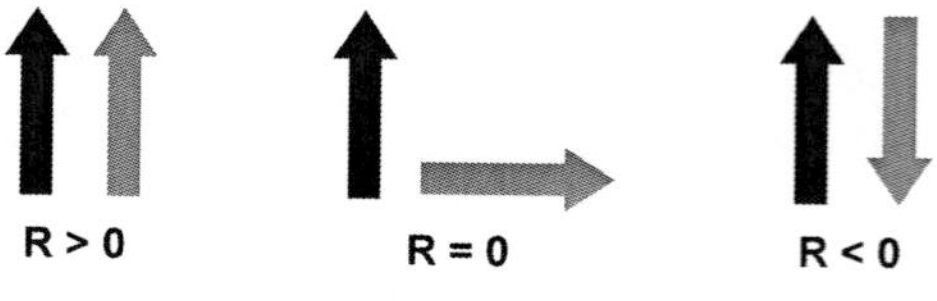

Figure 1.15 The rotational strength value is determined by the angle formed between the electric and magnetic dipole transition moments. The direction of the vector representation of $\boldsymbol{m}$ is such that, when viewed from its tip, the rotation of charge is anticlockwise, see Figure 2, while the direction of $\boldsymbol{\mu}$ follows that from the negative charge to the positive charge.

written as 0.927×10^{-38} cgs (centimeter-gram-second units). For this reason and to avoid the need to use numbers containing a power of 10, Moffitt introduced the reduced rotational strength $[R]$ as a dimensionless number:

$$[R] = 100R/\mu_D\mu_B = (1.08 \times 10^{-40})R \qquad (1.19)$$

Since the bands in a CD spectrum have positive and negative signs, in contrast with electronic absorption spectroscopy, the overall observed rotatary strength for molecules with groundstates which lack a permanent magnetic dipole moment should obey the "sum rule":

$$\sum R = 0 \qquad (1.20)$$

In many cases the observed bands will not fit this relationship, however, since significant CD intensity lies beyond the limits of what can be recorded by the available instrumentation. As will be described in greater detail in Chapter 3, the "sum rule" is particularly significant in the analysis of the CD intensity associated with exciton coupling.

1.2.2 Analysis of CD Spectra

CD spectra can provide important information about the absolute configuration of chiral molecules. The sign of the CD signal depends on the chromophore and the properties of the electronic transitions, so the analysis of CD spectra must always be made with reference to the corresponding electronic absorption spectra. As will be described in more detail in Chapter 7, with the recent advances in computer technology it is often possible to determine the absolute configuration of intrinsically chiral molecules by reproducing CD spectra using the standard calculation packages provided in commercially available DFT software packages. In other instances, however, the application of empirically derived rules or exciton coupling theory is more appropriate. The CD spectroscopy of organic molecules can be classified into four major groups. Each will be examined in greater detail in later chapters.

1. Symmetrical chromophores: Saturated carbonyls, double bonds without deformation and benzene chromophores belong to this group. The asymmetry produced by the rest of the molecule provides a sufficiently strong perturbation to the electronic structures that electronic transitions preferentially absorb left or right circularly polarised light. The resulting CD signal is generally not particularly strong. As will be described in greater detail in Chapter 2, empirical rules have been developed to analyse the CD spectra of certain groups of compounds which fit this category, but the application of these rules must be carried out carefully, since there are sometimes exceptions to the normal anticipated trend. In Chapter 5 the weak CD signals associated with magnetic dipole allowed $d \rightarrow d$ transitions of metal complexes will also be described,

2. Asymmetrical chromophores: This group consists of helicenes, skewed dienes, α,β- and β,γ-unsaturated carbonyls, biphenyls and compounds with deformed double bonds. Since the chromophores themselves are asymmetrical, the CD signals observed for this group are generally one order of magnitude larger than those observed in the context of symmetrical chromophores. As will be described in Chapter 7, the absolute structure of significantly twisted π-systems such as helicenes can be readily determined based on the use of molecular orbital (MO) calculations.
3. Exciton coupling: As will be described in greater detail in Chapters 3 and 5, when multiple asymmetrically orientated independent chromophores interact, there is scope for a Davydov splitting of the excited states. The CD signal associated with this mechanism is generally very strong, so the absolute configuration of molecules can be determined unambiguously and reliably using non-empirical approaches. The results obtained by this method have consistently been in accord with those obtained by the Bijvoet method in X-ray analyses. The most commonly studied systems contain two identical chromophores in close proximity, so it is important to understand the key concepts and terminology involved in this context. In the case of dimers, the two excited states produce CD spectra of opposite sign and equal intensity based on the "sum rule", equation (1.20). Figure 1.16 depicts the exciton coupling of a 1,2-diol *bis*(4-substituted benzoate), while Figure 1.17 provides details of how CD spectroscopy can be used to determine the configuration of the diol by studying the bisignate exciton couplet signal observed for the lowest energy band, which is polarised along the long axis of the benzoate chromophore almost parallel to the C–O bonds of the diol.[8] The amplitude of the signal, A, is defined as:

$$A = \Delta\varepsilon_1 - \Delta\varepsilon_2 \tag{1.21}$$

where $\Delta\varepsilon_1$ and $\Delta\varepsilon_2$ are the maximum intensities of the lower and higher energy bands, respectively. These are sometimes referred to as the first and second Cotton effects by organic chemists. The amplitudes of the exciton couplets are: (i) inversely proportional to the square of the interchromophoric distance, (ii) proportional to the extinction coefficient of their chromophores in the electronic absorption spectrum, and (iii) dependent on the interchromophoric projection angle of the edtms of the two chromophores, which are aligned almost parallel to the C–O bonds of the diol. There is a maximum at *ca.* 70° and minima at 0 and 180° when the edtms are coplanar, since there is no circulation of charge to generate a mdtm. The exciton couplets of systems are said to have positive chirality if they exhibit a plus-to-minus pattern in ascending energy terms in the CD spectrum and negative chirality if they exhibit a minus-to-plus pattern, Figure 1.17.
4. Induced circular dichroism: When an optically active substituent is attached to an achiral chromophore, an induced CD (ICD) signal is often observed that corresponds to the electronic absorption bands of the achiral chromophore. In Chapters 4 and 6 the analysis of ICD is

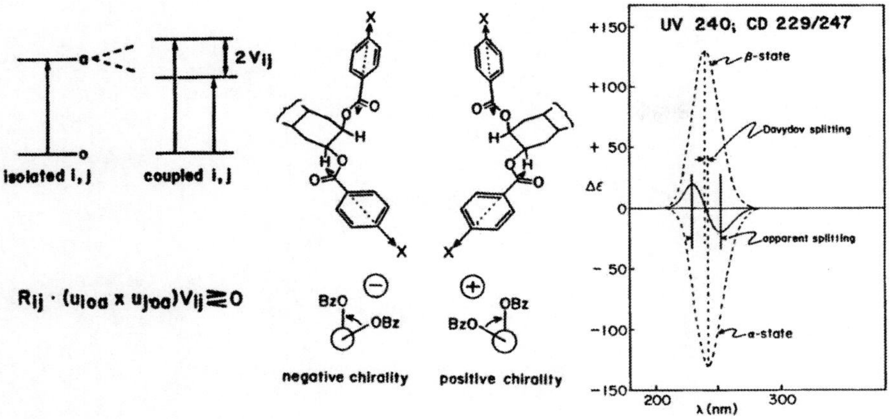

Figure 1.16 Exciton coupling between two identical 1,2-diol *bis*(4-substituted benzoate) chromophores (left). When viewed along the C–C bond of the 1,2-diol, molecules in which there is a clockwise twist arrangement in the alignment of the edtms of the first and second benzoate chromophores are defined as having positive excition chirality, while those with an anticlockwise twist arrangement are defined as having negative chirality (centre). The splitting of the two energy levels based on the Davydov splitting, $2V_{ij}$, results in a far larger apparent splitting of the CD spectrum (right). The minus-to-plus sequence in CD intensity in ascending energy terms is the exciton couplet signal that is observed for systems with negative chirality. The terminology is based on the sign observed at lower energy in the bisignate CD signal. [Reprinted with permission from *Pure Appl. Chem.*, **1984**, *56*, 1030. Copyright 1984, IUPAC.]

described in depth in the context of cyclodextrin inclusion compounds and binaphthyl-linked phthalocyanine complexes. Kuhn-Kirkwood coupled oscillator and "CD stealing" mechanisms from perturbation theory have to be taken into consideration in any theoretical description of the ICD of the electric dipole allowed transitions of the achiral chromophore.[9–11] The rotatory strength, R_1, associated with the Kuhn-Kirkwood coupled oscillator mechanism is given by:

$$R_1 = -K_1(\mathbf{r}_j - \mathbf{r}_i) \cdot (\boldsymbol{\mu}_j \times \boldsymbol{\mu}_i) \tag{1.22}$$

where $\mathbf{r}_j$ and $\mathbf{r}_i$ are the position vectors of the optically active and inactive chromophores, j and i, $K_1 = 2\pi v_i v_j V_{ij}/h(v_j^2 - v_i^2)$ and V_{ij} is the dipole–dipole interaction element which couples the two transitions:

$$\begin{aligned} V_{ij} = (1/4\pi\epsilon_0)\{\mathrm{Im}(\boldsymbol{\mu}_i \cdot \boldsymbol{\mu}_j/|\mathbf{r}_j - \mathbf{r}_i|^3) \\ - 3(\boldsymbol{\mu}_i \cdot (\mathbf{r}_j - \mathbf{r}_i))(\boldsymbol{\mu}_j \cdot (\mathbf{r}_j - \mathbf{r}_i))/|\mathbf{r}_j - \mathbf{r}_i|^5\} \end{aligned} \tag{1.23}$$

where ϵ_0 is the permittivity of a vacuum. In this mechanism, non-zero rotatory strength can arise if the alignments of the edtms on the chromophore and substituent are skewed.

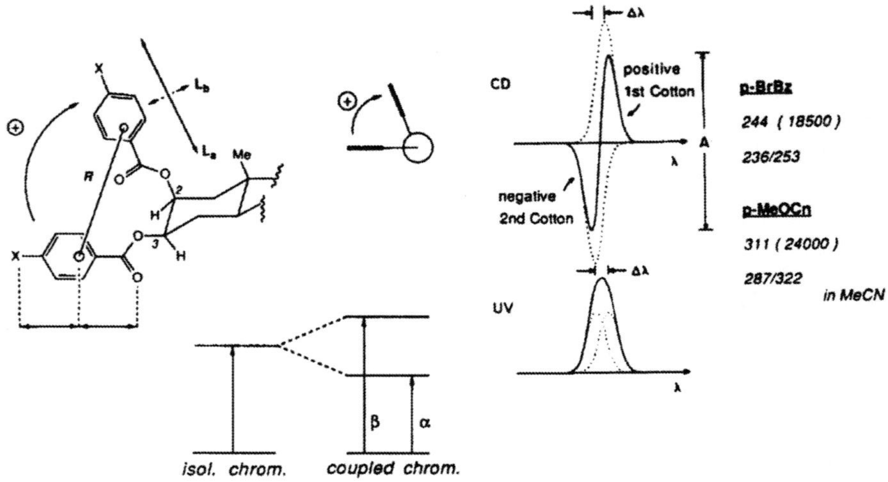

Figure 1.17 Typical patterns observed in the CD (right, top) and electronic absorption (right, bottom) spectra due to exciton coupling between two identical chromophores (bottom). The observed CD signal (solid line) is based on the differential absorbance of two component bands of opposite sign (denoted by dashed lines) separated by the Davydov splitting ($\Delta\lambda$). This usually results only in a broadening of the observed absorption band when the Davydov splitting is significantly smaller than the band widths of the two component bands (right). For systems with positive chirality (left) a plus-to-minus sequence is observed in ascending energy terms. A minus-to-plus sequence is observed for systems with negative chirality, see Figure 16. [Reprinted with permission from *Pure Appl. Chem.*, **1989**, *61*, 1193. Copyright 1989, IUPAC.]

The "CD stealing" mechanism directly transfers rotatory strength from a strongly optically active substituent to the central transition of the core π-system *via* dipolar coupling of the edtm of the achiral chromophore, *i*, and the mdtm of the optically active substituent, *j*. An example of this will be described in Chapter 6 in the context of binaphthyl-linked phthalocyanine complexes. The rotatory strength associated with this mechanism, R_2, is expressed as follows:

$$R_2 = -\mathrm{K}_2\mathrm{Im}(\boldsymbol{\mu}_\mathrm{i} \cdot \boldsymbol{m}_\mathrm{j}) \qquad (1.24)$$

where Im denotes the imaginary part and $\mathrm{K}_2 = 2\nu_\mathrm{i}V_\mathrm{ij}/\mathrm{h}(\nu_\mathrm{j}^2 - \nu_\mathrm{i}^2)$. This mechanism is also strongly geometry dependent and requires the transition moments at the two centres to be aligned. Although coupling between a mdtm of the central achiral chromophore and the edtm of the optically active substituent is also theoretically possible, electric dipole allowed bands tend to dominate the electronic absorption spectra, so coupling with the edtm will usually be the easier way to obtain an intense ICD signal, which can be readily used to derive structural information about a chiral system.

1.3 MCD Spectroscopy

In the 1840s, Michael Faraday observed that plane polarised beams of light are rotated during transmission through certain substances by the application of an applied magnetic field, due to the differential absorbance of lcp and rcp light which is induced by the applied field. In contrast with natural CD, which is relatively rare, the Faraday effect is a property of all matter. MCD spectroscopy, which is the spectroscopic method normally used to study magnetic optical activity, therefore differs significantly from CD spectroscopy, since a signal is observed even for compounds that lack a chiral centre. Prior to the emergence of MCD spectroscopy in the mid-1960s, after the development of Ti-Nb and Sn-Nb alloy superconducting magnet coils and the photoelastic modulators required to create an alternating beam of lcp and rcp light, magnetic optical rotatory dispersion (MORD) spectroscopy was the primary method used to study the Faraday effect. Although the MORD technique is now essentially obsolete, in the early 1960s MORD experiments demonstrated the potential utility of magnetic optical activity for studying the electronic structures of porphyrins, phthalocyanines, heme proteins and inorganic anions.[12] The MCD and MORD spectra are related by Kramers-Kronig transforms in the same manner as CD and ORD spectra.[13]

Over the last 40 years, MCD spectroscopy has provided many key insights into the optical spectroscopy and electronic structure of high symmetry molecules and complexes. The major difference between MCD spectroscopy, on the one hand, and the most widely used spectroscopic techniques that are based on an applied field, such as nuclear magnetic resonance (NMR) and electron paramagnetic resonance (EPR) spectroscopy, on the other, is that MCD spectroscopy is not based on resonance between spin states, but is instead based on the wavelength dependent absorption of circularly polarised light to form electronic excited states. The UV-visible absorption and MCD spectra of a molecular complex, therefore, contain the same set of spectral bands, but the band morphologies are different, owing to the effect of the applied magnetic field and the use of the same differential absorbance intensity scale that applies in CD spectroscopy. The five electronic quantum numbers ultimately form the theoretical basis for MCD spectroscopy, since the interaction of electronic states with an applied magnetic field is based upon magnetic dipole moments associated with the orbital and spin motion of the electrons of a compound or complex. The principal quantum number, n, was first proposed by Niels Bohr to account for the wave based properties of electron orbital motion. Subsequently, during the process of solving the Schrödinger wave equation for hydrogen, it was discovered that the orbital angular momentum (OAM) of the electron defined by the orbital quantum number, l, is spatially quantised in terms of the z-component, L_z, defined by the magnetic quantum number, m_l, Figure 1.18, based on precession at the Larmour frequency at fixed orientations around the axis of the applied magnetic field. In classical terms, circular motion of charge in a current loop has a magnetic dipole based on the current (I) and the area (A) enclosed by the orbital motion:

$$\boldsymbol{m} = IA = -\mathrm{e}/2\mathrm{m_e}L_z \tag{1.25}$$

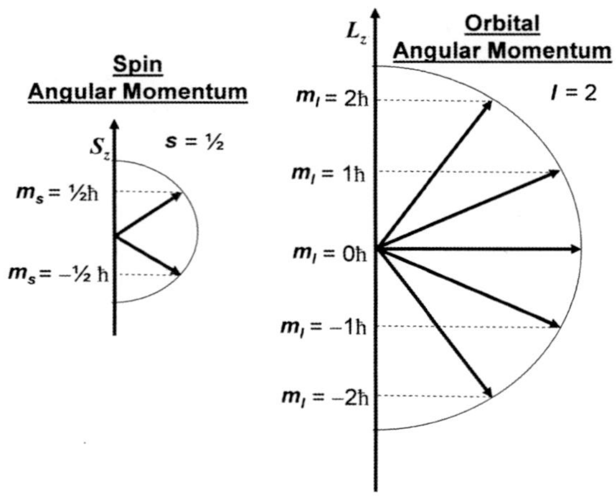

Figure 1.18 The origin of the l, m_l, s and m_s quantum numbers and the associated magnetic moments.

When the quantised nature of orbital angular momentum is factored in, the following equations can be derived:

$$\boldsymbol{m} = -\mathrm{e}/2\mathrm{m_e}\sqrt{l(l+l)}\hbar = \sqrt{l(l+l)}\mu_\mathrm{B} \tag{1.26}$$

$$\boldsymbol{m}_z = -m_l\mu_\mathrm{B} \tag{1.27}$$

where $l=0, 1\ldots n\text{–}1$ and $m_l = l, l-1\ldots0, -l+1, -1$, and e and $\mathrm{m_e}$ are the charge and mass of the electron, respectively, ħ is the Dirac constant (h/2π) and μ_B is the Bohr magneton. The negative sign arises because magnetism is still defined by physicists in terms of a positive charge carrier, for historical reasons.

In the case of spin angular momentum, the spin properties of an electron are quantised so that $s=1/2$ and $m_s=\pm1/2$, Figure 1.18. Electron spin is more efficient at generating a magnetic moment than the orbital motion so the magnetic moment equation for S_z contains the Landé factor (g-factor), which has a value of 2.0023:

$$\boldsymbol{m}_z = -g_L m_s \mu_\mathrm{B} \approx -\mu_\mathrm{B} \tag{1.28}$$

which was demonstrated empirically during early spectroscopic studies to be:

$$g_L = 1 + \frac{J(J+1) + S(S+1) - L(L+1)}{2J(J+1)} \tag{1.29}$$

Since the g-factor associated with orbital angular momentum has a value of 1 when $S=0$, this factor does not have to be included during the calculation of the Zeeman splitting of states for most organic molecules and the aromatic and heteroaromatic complexes of diamagnetic main group and closed shell d^{10} metal ions. It is a significant factor, however, with paramagnetic transition metal ions and, as will be described in Section 1.3.4, this can greatly complicate the theoretical treatment required to analyse an MCD spectrum.

When these atomic quantum number properties are introduced into the electronic description of a molecule, any calculation of the optical spectrum requires a solution of the Schrödinger wave equation both in terms of the states (Ψ_A) and energy levels (E_A) of the system and of the Hamiltonian (H):

$$H\Psi_A = E_A\Psi_A \tag{1.30}$$

since this allows calculation of the experimentally observable transition energy and oscillator strength values (M^{AJ}), which are associated with the transition dipole moment between the ground and excited states, A and J:

$$M^{AJ} = \langle \Psi_A | \boldsymbol{\mu} | \Psi_J \rangle \tag{1.31}$$

Introductory treatments of the selection rules for optical spectroscopy often focus primarily on whether a transition is spin ($\Delta S = 0$) and/or Laporte ($\Delta l = \pm 1$) forbidden, which in the case of centrosymmetric systems can also be expressed in terms of a change in parity such that u → u and g → g transitions are forbidden. An additional selection rule, which plays a key role in MCD spectroscopy, that only $\Delta M_L = 0, \pm 1$ transitions are fully allowed is often overlooked (it should be noted that in certain circumstances this is also an oversimplification), since it is difficult to derive information based on this selection rule from standard UV–visible absorption spectra. An incident photon can provide a maximum of one quantum of orbital angular momentum ($\Delta M_L = \pm 1$), since the electric and magnetic vectors of the electromagnetic wave can rotate a maximum of once per wavelength, see Figure 1.12, and the total angular momentum within an isolated system must be conserved in both magnitude and direction according to Newton's third law of motion. For example, the absorption of a photon by a heteroaromatic π-system results in both a linear and a circular redistribution of charge within the xy-plane, see Figure 1.3. Although strong magnetic dipoles are, therefore, induced in the excited states there is no net CD signal since in the absence of a chiral centre the absorbance of lcp and rcp light is equally likely.

1.3.1 Zeeman Splitting of States

The role of the magnetic vector of electromagnetic radiation tends to be neglected within introductory treatments of optical spectroscopy with the exception of descriptions of line spectra arising from isolated atoms in the gas phase and in astronomical observations. In introductory physical chemistry courses, this phenomenon is sometimes explained no further than the normal Zeeman effect observed during relaxation of an atomic hydrogen $1p$ orbital electron to the $1s$ orbital. The emission band is described as splitting into parallel (||) and perpendicular (⊥) plane polarised bands, Figure 1.19, based on

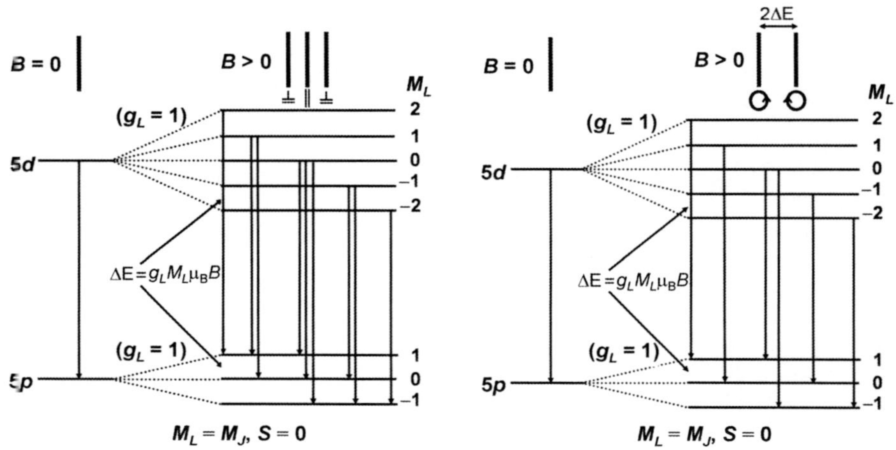

Figure 1.19 The normal Zeeman effect. The emission from an atomic lamp source placed within a magnet observed perpendicular (left) and parallel (right) to the lines of force. It should be noted that when $S=0$ and the Russell–Saunders spin coupling mechanism is applicable, $J=L$ and $M_J=M_L$ as shown here.

$\Delta m_l=0$ and $\Delta m_l=\pm1$ selection rules, respectively, when the field lines are applied perpendicular to the emitted beam. What is seldom mentioned is that Pieter Zeeman's experiments (for which he shared the Nobel prize for physics in 1902) involved emission from a Cd 5*d* orbital to a 5*p* orbital with the magnetic field lines applied in ***both*** a parallel and a perpendicular arrangement relative to the axis of light propagation. In Zeeman's words during his Nobel prize lecture:[14] "*....But let us first consider the rays which run parallel with the lines of force.... The opposite circular oscillations of the electrons excite two circularly polarised rays rotating in opposite directions, one having a longer and the other a shorter period of oscillation than the original spectral line. The original spectral line splits up under the action of the magnetic field into two components which are circularly polarised in opposite directions. The light source emits two-colour light....*".

The MCD technique would probably be more widely appreciated if Zeeman's results were explained with greater emphasis placed on this first ever example of a magnetic circularly polarised emission spectrum. The MCD experiment in many respects is simply the reverse of Zeeman's observation, in that differential absorbance of lcp and rcp photons is induced by the Zeeman splitting of states within an applied magnetic field.

Since cadmium has a closed shell d^{10} configuration, spin orbital angular momentum is not a factor in the normal Zeeman effect. When Zeeman studied metals with unpaired electrons such as sodium, however, multiple emission lines were observed in both the presence and the absence of an external magnetic field. This is usually referred to as the anomalous Zeeman effect. In the case of sodium, a doublet is observed for the 3*p* → 3*s* emission at zero

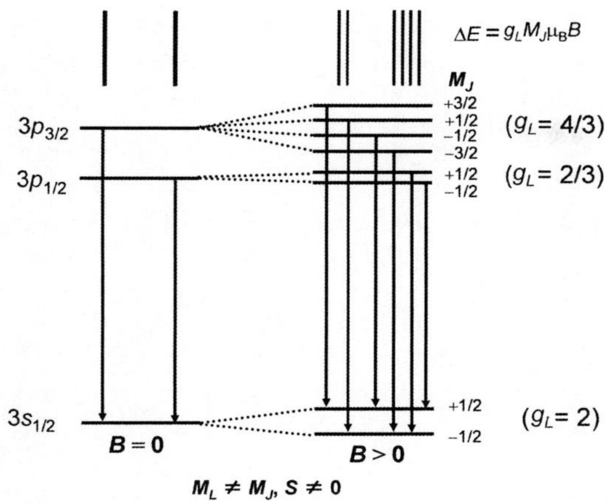

Figure 1.20 The anomalous Zeeman effect. The emission from an atomic lamp source placed within a magnet observed parallel to the lines of force. Unlike the case in Figure 1.19, J and M_J are now used since $S \neq 0$.

field, since the electron spin angular momentum can be aligned either with or against the strong internal magnetic dipole created by the orbital motion of the unpaired electron. The anomalous Zeeman effect represents direct spectral evidence that spin and orbital angular momenta of the individual electrons couple to provide the total angular momentum quantum numbers, J and M_J, for the electronic state based on the L, M_L, S and M_S quantum numbers. In the case of lighter atoms (Z <40), spin–orbit coupling is relatively weak and the L and S quantum numbers can be combined according to the Russell–Saunders coupling scheme (as opposed to the strong-field *j–j* coupling scheme observed in the case of heavier atoms where the spin–orbit coupling is stronger) to give the total angular momentum of the state J ($J = L + S$). When a magnetic field is applied, there is a lifting of state degeneracies into $2J + 1$ microstates, based on the M_J quantum number for orientation relative to the applied field. The Zeeman splitting of the states, Figure 1.20, can then be defined:

$$\Delta E = \frac{e}{2m_e}(L_z + 2S_z)B = g_L \mu_B M_J B \tag{1.32}$$

The factor of 2 for S_z is related to g_L, equation (1.29). The value for the z-component of the magnetic moment aligned with the axis of light propagation can be defined as:

$$\boldsymbol{m}_z = -\mu_B(L_z + 2S_z) \tag{1.33}$$

The absorption of a chiral lcp or rcp photon results in either a left- or right-handed redistribution of charge, since the total OAM within the system must be conserved. The band morphologies observed in an MCD spectrum can provide key information about the Zeeman splittings of the ground and/or excited states of high symmetry compounds and complexes and the signal intensity can sometimes be used to derive quantitative values for the ground and/or excited state magnetic dipole moments.

1.3.2 MCD Intensity Mechanism, Sign Conventions and Intensity Units

The modern theoretical background to MCD spectroscopy is largely attributable to Buckingham, Stephens and Schatz.[2,15] Although electronic absorption and MCD spectral data are normally plotted in the literature in terms of extinction coefficients ε (M^{-1} cm^{-1}) and $\Delta\varepsilon_M$ (M^{-1} cm^{-1} T^{-1}) based on Beer–Lambert's law, the quantitative analysis of MCD spectra is usually based on an analysis of the component spectral bands and the estimation of the magnitudes of the three Faraday terms, $\mathcal{A}_1$, $\mathcal{B}_0$ and $\mathcal{C}_0$, which were originally defined within the context of MORD spectroscopy based on a different set of definitions referred to as A, B and C.[9,16] For a fully allowed electronic transition using the Rigid–Shift, Born–Oppenheimer and Franck–Condon approximations, the MCD intensity equation, according to the modified conventions recommended by Stephens, Piepho and Schatz,[2,15] is:

$$\frac{\Delta A_{l-r}}{\mathrm{E}} = 152.5Bcl[\mathcal{A}_1\left(-\frac{\mathrm{df}}{\mathrm{dE}}\right) + (\mathcal{B}_0 + \mathcal{C}_0/\mathrm{k}T)\mathrm{f}] \tag{1.34}$$

ΔA_{l-r} is the differential absorbance of lcp and rcp light, B is the field strength, cl is the product of the concentration (mol. l^{-1}) and pathlength (cm), E represents the energy coordinate in cm^{-1} and is used here to signify that the expression is for the entire spectral band rather than for just one wavelength, and f is a normalised band shape function (normally a Gaussian-shaped curve). As will be described in more detail below, the $\mathcal{A}_1$, $\mathcal{B}_0$ and $\mathcal{C}_0$ terms provide information, respectively, about the Zeeman splitting of the absorption bands for lcp and rcp light, the field-induced mixing of zero-field states, and the Zeeman splitting based groundstate population adjustment.

The dipole strength of the absorption intensity,[1] in the absence of an applied field, is:

$$\mathcal{D}_0 = \frac{3.062 \times 10^{-3}}{\nu_{\max}} \int \varepsilon \mathrm{d}\nu \approx \frac{\langle A \rangle_0}{326.6cl} \approx \frac{\langle \varepsilon \rangle_0}{326.6} \tag{1.35}$$

where $\mathcal{D}_0$ is expressed in (Debye)^2 units (D^2),[1,17] ν_{max} is the band centre energy in cm^{-1} and $\langle A \rangle_0$ is the zeroth moment of the intensity for the entire absorption band. The right-hand side equation is valid for any symmetrical band shape. It is important to note that throughout this section energies are in cm^{-1} and field strengths are in Tesla (T).

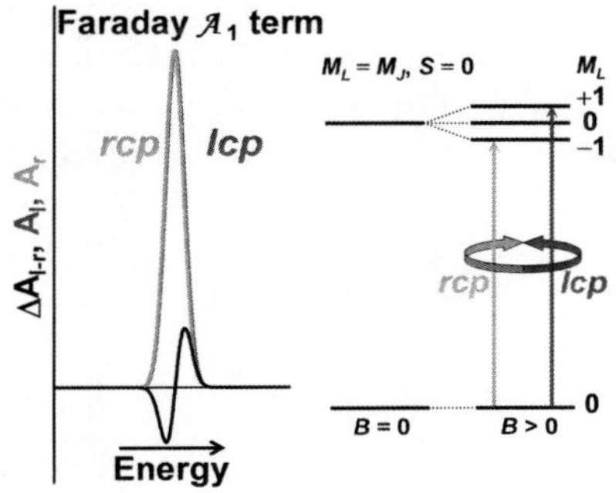

Figure 1.21 Selection rules for the absorption of circularly polarised light based on a transition between an orbitally nondegenerate groundstate and an orbitally degenerate excited state within a magnetic field applied parallel to the axis of light propagation. A derivative-shaped Faraday $\mathcal{A}_1$ term is observed due to the Zeeman splitting of states and the associated energy separation of bands absorbing lcp and rcp light based on the $\Delta M_L = \pm 1$ selection rule. In this example, since $S = 0$ for closed shell atoms or molecules, $J = L$ and $M_J = M_L$.

MCD spectra are recorded by mounting a magnet into the sample compartment of a CD spectrometer, in the Faraday alignment. A parallel applied field is often referred to as the Faraday alignment because it formed the basis of Michael Faraday's experiments in the 1840s. Since a perpendicular applied field tended to be used within Zeeman spectroscopy (*i.e.* excited state emission experiments), it is sometimes referred to as the Zeeman magnet arrangement. An electromagnet with a field strength of 1–2 T is usually sufficient for studying Faraday $\mathcal{A}_1$ or $\mathcal{B}_0$ terms, while superconducting magnets with field strengths of 5–7 T are often used to study the temperature and field dependence properties of Faraday $\mathcal{C}_0$ terms.

The $\mathcal{A}_1$ term, Figure 1.21, arises from the Zeeman splitting of an orbitally degenerate excited state. There is a highly distinctive first derivative band shape due to the separation of the band centres for the absorption of lcp and rcp light. A consideration of equation (1.36) makes it clear that the $\mathcal{A}_1$ term intensity is based on the relative magnitudes of the combined orbital (L_z) and spin (S_z) angular momenta of electrons within the ground and excited states, labelled as A and J, respectively:

$$\mathcal{A}_1 = \frac{1}{d_A}\sum \langle \mathrm{J}|\boldsymbol{m}_z|\mathrm{J}\rangle - \langle \mathrm{A}|\boldsymbol{m}_z|\mathrm{A}\rangle \left([\boldsymbol{\mu}_{-1}^{\mathrm{AJ}}]^2 - [\boldsymbol{\mu}_{+1}^{\mathrm{AJ}}]^2\right) \tag{1.36}$$

where $[\boldsymbol{\mu}_{-1}^{\mathrm{AJ}}]$ and $[\boldsymbol{\mu}_{+1}^{\mathrm{AJ}}]$ are the electric dipole matrix elements associated with the absorption of lcp and rcp light, respectively, while d_{A} relates to the degree of degeneracy of the excited states. It should be noted that the sign convention for handedness used in the subscripts of the matrix elements is the opposite of that used in ΔA_{l-r}. In the case of aromatic π-systems, absorption of a chiral lcp or rcp photon results in either a left- or right-handed redistribution of charge, since the total orbital angular momentum within the system must be conserved. The magnetic field lines induced by the circular redistribution of charge all point along the *z*-axis at the centre of the cyclic polyene, creating a magnetic dipole moment perpendicular to the plane of the π-system. Within the MCD experiment the splitting of the optically accessible microstates of an orbitally degenerate excited state is:

$$\Delta E = 2g_L\mu_B M_J B \tag{1.37}$$

The factor of 2 is based on the fact that the microstate accessed by a photon of one handedness will be stabilised due to an alignment of the magnetic dipole moment with the applied field, while that accessed by a photon of the opposing handedness will be destabilised since the magnetic dipole is aligned against the field, Figure 1.22.

The Gaussian-shaped $\mathcal{B}_0$ term arises from second order effects based on the field induced mixing of the zero-field states *via* mdtms. $\mathcal{B}_0$ terms completely dominate the MCD spectrum when there is no three-fold or higher rotation axis, since there are no orbitally degenerate states that can be split due to the Zeeman effect. $\mathcal{B}_0$ terms are also present in the spectra of high symmetry complexes but tend to be significantly less intense than the $\mathcal{A}_1$ and $\mathcal{C}_0$ terms,

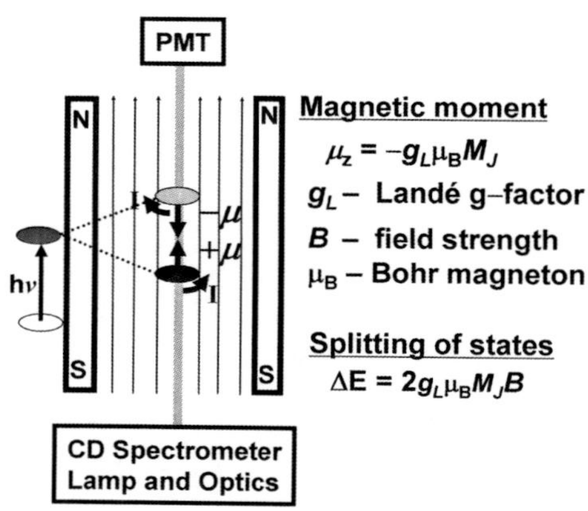

Figure 1.22 The direction of the induced magnetic moments within a cyclic polyene excited state can be derived from Ampere's rule (right-hand rule) for a solenoid (*i.e.* for current in a loop). When the fingers of the right hand are curled in the flow direction for conventional current flow (*i.e.* for a positive charge carrier) the induced magnetic dipole moment is aligned in the direction the thumb is pointing. The MCD magnet's sample horizontal bore is surrounded by a solenoid to generate a high magnetic field, typically of 1–7 T, with field lines running from the south to north parallel to the axis of light propagation in the sample compartment of the CD spectrometer.

since there is a dependence on the energy separation of states J and K (ΔE_{KJ}) which are mixed by the applied field. The $\mathcal{B}_0$ term intensity, in the absence of significant field induced mixing between the ground and excited states, is:

$$\mathcal{B}_0 = \frac{2}{d_A}\Re e \sum \frac{\langle J|\boldsymbol{m}_z|K\rangle}{\Delta E_{KJ}}\left([\boldsymbol{\mu}_{-1}^{AJ}][\boldsymbol{\mu}_{+1}^{KA}] - [\boldsymbol{\mu}_{+1}^{AJ}][\boldsymbol{\mu}_{-1}^{KA}]\right) \tag{1.38}$$

When a structural modification results in a small zero–field splitting, relative to the spectral bandwidth, of what would otherwise be an orbitally degenerate ππ* excited state, the resulting derivative-shaped signal is often referred to as a pseudo- $\mathcal{A}_1$ term, Figure 1.23.

The $\mathcal{C}_0$ term arises, Figure 1.24, from the Boltzmann population distribution across a degenerate groundstate and, therefore, exhibits strong $1/kT$ temperature dependence, equation (1.39). $\mathcal{C}_0$ terms are usually associated with the presence of an open shell transition metal. When the splitting of the microstates is small relative to the spectral bandwidths of the lcp and rcp light-specific absorption bands, the $\mathcal{C}_0$ term is essentially Gaussian shaped. The $\mathcal{B}_0$ terms can be readily distinguished from $\mathcal{C}_0$ terms by measuring spectra at both room and cryogenic temperatures since they lack $1/kT$ temperature dependence.

$$\mathcal{C}_0 = -\frac{1}{d_A}\sum \langle A|\boldsymbol{m}_z|A\rangle\left([\boldsymbol{\mu}_{-1}^{AJ}]^2 - [\boldsymbol{\mu}_{+1}^{AJ}]^2\right) \tag{1.39}$$

Unfortunately the legacy of MORD spectroscopy has led to confusion within the literature as to how to refer to the signs of the three Faraday terms. Several different sets of intensity mechanism equations have been published and the most recent equations recommended by Stephens[15a] and Piepho and Schatz,[2] which are used in this book, have not been universally adopted by all of the research groups active in the field of MCD spectroscopy. An earlier set of

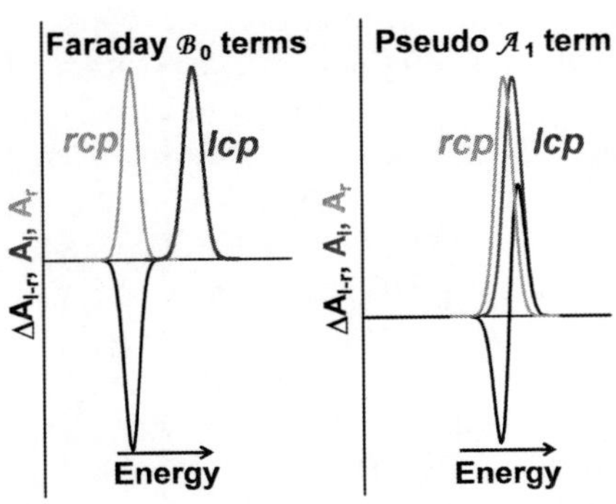

Figure 1.23 The appearance and the intensity of coupled Faraday $\mathcal{B}_0$ terms depend upon the extent of the zero-field splitting of states relative to the bandwidth. The splitting of the positively and negatively signed and Gaussian-shaped MCD signals (black) is determined by the energy separation of the lcp and rcp light absorption bands (dark and light grey) that arise within the applied field due to field induced mixing of states via a mdtm.

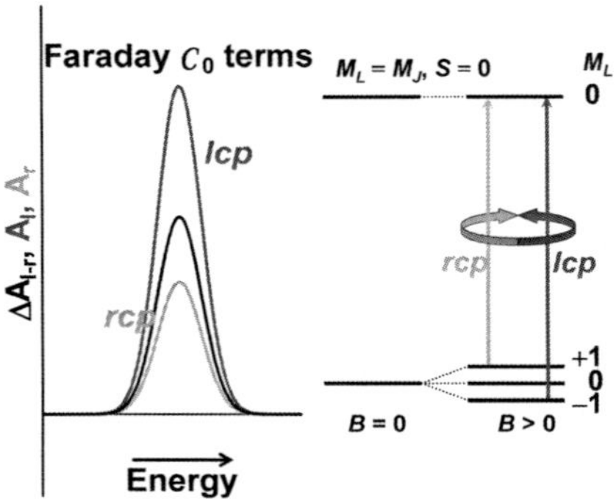

Figure 1.24 Selection rules (right) for the absorption of circularly polarised light based on a transition between an orbitally degenerate groundstate and an orbitally nondegenerate excited state within a magnetic field applied parallel to the axis of light propagation. The $\mathcal{C}_0$ terms are extremely temperature sensitive. The $\mathcal{C}_0$ term has a $1/kT$ temperature dependence since the groundstate microstates are populated according to a Boltzmann distribution. At very low temperatures, a symmetric Gaussian-shaped Faraday $\mathcal{C}_0$ term is observed because only the lowest split state contributes to the MCD intensity. Understandably, the intensity of this $\mathcal{C}_0$ term dominates the MCD spectrum at low temperatures. At temperatures above the temperature that the lowest state exists for 100% of the population, the measured $\mathcal{C}_0$ term shape begins to become asymmetric, due to the separation of the band centres of transitions absorbing lcp and rcp light, until both states are equally populated at very high temperatures, at which point an $\mathcal{A}_1$-term-like derivative-shaped band will be observed. In this example $S=0$ so $M_L = M_J$. The Faraday $\mathcal{C}_0$ terms that arise when $S \neq 0$ are described in Figures 1.26 and 1.27.

definitions for the three Faraday terms, which were adopted in the mid-1960s,[18–20] usually denoted by *A*, *B* and *C*, have also been used extensively. Within MORD spectroscopy, Figure 1.25, a positive Verdet constant and ϕ value for rotation of a plane polarised beam correspond to clockwise right-handed rotation when viewed towards the source. Absorption of incident lcp light, therefore, results in a positive signal within the CD intensity definition, but a negative Verdet constant based on negatively signed component Faraday *A*, *B* and/or *C* terms. When the MORD definitions for the Faraday terms were initially adopted in MCD spectroscopy, this resulted in Gaussian-shaped MCD bands with a negative ΔA_{l-r} CD signal initially being referred to as positive *B* or *C* terms by researchers such as Michl[18] and Djerassi[19] in key early landmark papers on MCD spectroscopy. Research groups who have subsequently adopted the modified moments definitions of Stephens[15a] and Piepho and Schatz[2] are now able to avoid this source of confusion and are able to refer to a negative ΔA_{l-r} signal as a negative Faraday $\mathcal{B}_0$ term, since the sign of $\mathcal{B}_0$ matches that of ΔA_{l-r}. It should be noted, however, that the sign convention for the $\mathcal{A}_1$ term is the same as that of the earlier definition of *A*, since the negative and positive *A* term terminology, based on the high energy lobe of the

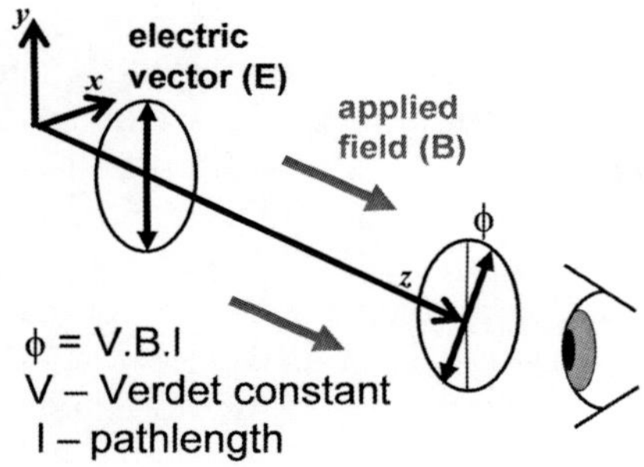

Figure 1.25 The Faraday effect within MORD spectroscopy. A plane polarised beam is rotated by a magnetic field applied parallel to the axis of propagation. The rotation is defined as positive if the beam is rotated to the right in a clockwise direction when viewed towards the source.

derivative-shaped signal, was well established and did not cause the same level of confusion as the Gaussian-shaped *B* and *C* term definitions. In summary, in describing an MCD spectrum we recommend the use of the following style, "positive $\mathcal{A}_1$ term" for a derivative-shaped signal that is positive at the high energy side of the cross-over point and "negative $\mathcal{B}_0$ term" and/or "negative $\mathcal{C}_0$ term" for Gaussian-shaped bands that have a negative ΔA_{l-r} signal based on equations (1.34–1.39) using the modified conventions recommended by Stephens, Piepho and Schatz.[2,15]

Another complication faced when comparing MCD spectra that have been reported in the literature is that there are two different sets of units currently in widespread use to express MCD intensity. For historical reasons some research groups continue to use a field normalised, molar ellipticity unit, $[\theta]_M$. Since θ is proportional to the differential absorbance of lcp and rcp light, however, ΔA_{l-r} has increasingly been adopted, and $\Delta\varepsilon_M$ based on the extinction coefficient for a 1 M solution at a field strength of 1 T now tends to be the preferred intensity unit for MCD spectroscopy. The conversion factor is $[\theta]_M = 3298\Delta\varepsilon_M$, where $[\theta]_M$ is expressed in units of deg.cm^2.dmol^{-1}. T^{-1}.

1.3.3 Quantitative Analysis of Groundstate and Excited State Magnetic Dipole Moments

Comparison of equations (1.36), (1.38) and (1.39) for $\mathcal{A}_1$, $\mathcal{B}_0$ and $\mathcal{C}_0$ with equation (1.40) for the dipole strength of the absorption band, $\mathcal{D}_0$, demonstrates that the electric dipole contribution cancels out in $\mathcal{A}_1/\mathcal{D}_0$, $\mathcal{B}_0/\mathcal{D}_0$ and $\mathcal{C}_0/\mathcal{D}_0$ ratios:

$$\mathcal{D}_0 = \frac{1}{d_A}\sum \left[\boldsymbol{\mu}_{\pm1}^{AJ}\right]^2 \qquad (1.40)$$

Since $\mathcal{D}_0$ can be readily derived from the corresponding UV–visible absorption spectra, the MCD technique can be used to directly quantify the magnetic moments associated with the ground and excited states based on the $\mathcal{C}_0/\mathcal{D}_0$ and $\mathcal{A}_1/\mathcal{D}_0$ ratios, respectively.[2,15a] Method of moments analysis was introduced to MCD spectroscopy by Henry, Schnatterly and Slichter,[17] and has since been used extensively to derive the magnetic moments of the ground and/or excited states from MCD spectral data. In many instances this is not practical, however, since there is extensive overlap between close-lying spectral bands, and spectral band deconvolution analyses have to be attempted instead.

The $\mathcal{A}_1$, $\mathcal{B}_0$, $\mathcal{C}_0$ and $\mathcal{D}_0$ definitions, which we strongly recommend, have not been universally adopted by all of the active MCD research groups and two older sets of definitions can still be found in recent MCD papers. Ratio values based on A, B and C definitions, which were originally derived in the 1930s within the calculation of the Verdet constant,[9d,16] are sometimes still encountered, as are ratio values derived from a later set of definitions, denoted here as $\mathcal{A}_{1(\mathrm{OLD})}$, $\mathcal{B}_{0(\mathrm{OLD})}$, $\mathcal{C}_{0(\mathrm{OLD})}$, which were used briefly by Stephens in the early 1970s in the context of MCD spectroscopy.[21] It should be noted that Stephens subsequently recommended the use of the $\mathcal{A}_1$, $\mathcal{B}_0$, $\mathcal{C}_0$ and $\mathcal{D}_0$ definitions in equations (1.34–1.39) instead.[2,15a] Within any review of the MCD literature, the conventions used, therefore, need to be carefully checked. The conversion factors to derive $\mathcal{A}_1$, $\mathcal{B}_0$ and $\mathcal{C}_0$ values from the older definition are:

$$\mathcal{A}_1 = -\frac{2}{3}, \mathcal{A}_{1(\mathrm{OLD})} = \frac{2}{3}A \tag{1.41}$$

$$\mathcal{B}_0 = -\frac{2}{3}\ \mathcal{B}_{0(\mathrm{OLD})} = -\frac{2}{3}B \tag{1.42}$$

$$\mathcal{C}_0 = -\frac{2}{3}\ \mathcal{C}_{0(\mathrm{OLD})} = -\frac{2}{3}C \tag{1.43}$$

$$\mathcal{D}_0 = \frac{1}{3}\ \mathcal{D}_{0(\mathrm{OLD})} = \frac{1}{3}D \tag{1.44}$$

A set of conversion factors quoted by Piepho and Schatz[2] has caused considerable confusion for experimentalists. An extra factor of $1/\mu_B$ is included in this context, which while technically correct in terms of their starting equations is not applicable for $\mathcal{A}_1$, $\mathcal{B}_0$, $\mathcal{C}_0$ and $\mathcal{D}_0$ values derived from observed ΔA and A values on the basis of equations (1.34) and (1.35), since μ_B is already included within $\mathcal{A}_1/\mathcal{D}_0$, $\mathcal{B}_0/\mathcal{D}_0$ and $\mathcal{C}_0/\mathcal{D}_0$ ratio calculations as a factor of 152.5/326.6. The units for $\mathcal{A}_1$, $\mathcal{B}_0$, $\mathcal{C}_0$ derived from a direct application of these equations to experimental spectral data are $D^2.\beta$ for $\mathcal{A}_1$ and $D^2.cm.\beta$ for $\mathcal{B}_0$ and $\mathcal{C}_0$, therefore, as was the case previously with A, B and C.

The factor of 2 in equations (1.41–1.43) ensures that the z-component of the magnetic dipole moment for the ground and excited states can be derived directly from the $\mathcal{A}_1/\mathcal{D}_0$ and $\mathcal{C}_0/\mathcal{D}_0$ ratio values in instances where $S=0$ and $g_L=1$:

$$\boldsymbol{m}_z(\text{excitedstate}) = -\mathcal{A}_1/\mathcal{D}_0 \qquad \boldsymbol{m}_z(\text{groundstate}) = \mathcal{C}_0/\mathcal{D}_0 \tag{1.45}$$

The factor of 1/3 is related to the presence or absence of orientation averaging, which takes into account the fact that, in solution, molecules are randomly orientated so that the molecular axis does not coincide with that of the applied field. In practical terms, although this can affect the magnitude of $\mathcal{A}_1$, $\mathcal{B}_0$, $\mathcal{C}_0$ and $\mathcal{D}_0$, it cancels out within $\mathcal{A}_1/\mathcal{D}_0$, $\mathcal{B}_0/\mathcal{D}_0$ and $\mathcal{C}_0/\mathcal{D}_0$ ratio calculations. Random orientation is sometimes denoted with overbars. It should be noted that the difference between randomly orientated solution spectra and spectra measured for a specific molecular orientation within a single crystal or an argon matrix can potentially be a significant factor for complexes with anisotropic symmetries, and in those circumstances the equations required to analyse the Faraday terms may be significantly more complex than those described here.[2]

Even in the absence of methods of moments analysis, approximate values for the key parameters can still be derived by considering the extrema of the recorded MCD curves. In cases where the observed signal is derivative shaped, *i.e.* the equations hold when the $\mathcal{A}_1/\mathcal{B}_0$ ratio is greater than or at least equal to 0.5Γ (Γ: the width at half-maximum absorption), if $\Delta\varepsilon_{M1}$ and $\Delta\varepsilon_{M2}$ are the maximum and minimum intensity values *including their sign* for the peaks at longer wavelength (λ_1) and shorter wavelength (λ_2) and (Sum) and (Dif) stand for their sum and difference, respectively, Briat[22] demonstrated that:

$$\mathcal{A}_1\mu_B \sim -\Delta(\text{Dif})/27.4\lambda_0\varepsilon_M \quad (1.46)$$

$$[\mathcal{B}_0+(\mathcal{C}_0/kT)]\mu_B \sim \Delta(\text{Sum})/15.8\lambda_0 \quad (1.47)$$

where Δ (bandwidth) and λ (wavelength of maximum absorption) are expressed in the same units (usually nanometers) and Γ is expressed in cm^{-1}. The dipole strength, $\mathcal{D}_0$, of the relevant transition can be estimated from $\mathcal{D}_0 \sim 3.07 \times 10^{-3}\varepsilon_M\Delta/\lambda_0$ where ε_M is the maximum molar extinction coefficient. It then follows that $\mathcal{A}_1/\mathcal{D}_0$ and $[\mathcal{B}_0 + (\mathcal{C}_0/kT)]/\mathcal{D}_0$ can be expressed more simply as:

$$\mathcal{A}_1/\mathcal{D}_0 \sim -1.3\Gamma(\text{Dif})/\varepsilon_M \quad (1.48)$$

$$[\mathcal{B}_0+(\mathcal{C}_0/kT)]/\mathcal{D}_0 \sim 2.28((\text{Sum})/\varepsilon_M) \quad (1.49)$$

to derive values comparable to those obtained from equations (1.34) and (1.35). In cases where a derivative-shaped $\mathcal{A}_1$ signal cannot be identified clearly in the spectrum and a Gaussian-shaped curve is instead observed, Briat demonstrated that the following relationship holds:

$$[\mathcal{B}_0+(\mathcal{C}_0/kT)]/\mathcal{D}_0 \sim 3.38\Delta\varepsilon_{M\max}/\varepsilon_M \quad (1.50)$$

where $\Delta\varepsilon_{M\max}$ represents the maximum value of $\Delta\varepsilon_M$, this maximum occurring at a wavelength very close to or coincident with that of the absorption maximum. In spectra where the most intense spectral bands do not overlap to any significant extent, the results obtained using Briat's approach have been found to provide values that are very close to those obtained from method of moments analyses. In most instances, however, where there is significant overlap between close-lying intense spectral bands, a more quantitative approach based on spectral band deconvolution should be adopted.

1.3.4 MCD Spectroscopy of Transition Metal Complexes

Since the mid-1960s, MCD spectroscopy has been used extensively in biochemistry for studying haem proteins, such as myoglobin and haemoglobin,[23] and other biologically significant transition metal complexes. The complex band morphologies arising from the three Faraday terms result in major changes within the overall MCD spectrum when the oxidation or spin state of the metal ion is changed. In contrast, only relatively minor changes in band centres and intensities are typically observed in the UV–visible absorption spectra. Extensive studies of Faraday $\mathcal{C}_0$ term intensity at cryogenic temperatures have provided key ground and excited state degeneracy information that cannot easily be derived from the analysis of UV–visible absorption spectra and theoretical calculations, alone. The MCD technique is complementary to EPR spectroscopy in this regard, since that technique provides key information about the electron spin state of paramagnetic complexes.

The redox chemistry of transition metal complexes is significantly more complex than that of main group metal complexes owing to the partially filled *d* shell. Ligand to metal and metal to ligand charge transfer (LMCT, MLCT) bands are typically observed in both the absorption and MCD spectra in addition to the $\pi \rightarrow \pi^*$ bands. The ligand field splitting of the *d* orbitals in haems tends to be essentially tetragonal, on the basis of the tetradentate nature of the protoporphyrin IX ligand and some form of axial ligation from above and below the ring. In an O_h symmetry environment the ligand field splitting (Δ_{OCT}) is typically of the order of 10 000 cm^{-1}. Most six-coordinate D_{4h} symmetry complexes are axially distorted ($x = y \neq z$) so that the degeneracy of the e_g and t_{2g} levels is lifted, while lower symmetry complexes contain a rhombic distortion ($x \neq y \neq z$) in which there is a complete lifting of *d* orbital degeneracy. In the case of Fe(II) and Fe(III) complexes, the spin state of the central metal is usually determined by the nature of axial ligation from above and below the ring. Strong ligands such as CO and CN^-, which act as π-acceptors based on back bonding, result in low spin complexes by increasing the separation of the e_g and t_{2g} orbitals, while weak ligands, which act as π-donors, such as F^-, reduce the separation and result in a high spin state with $S > 1/2$. In the case of Fe(III) haems, ligation with azide (N_3^-) can also result in an intermediate spin state with $S = 3/2$. In complexes where there are unpaired electrons so that $S > 0$, spin–orbit coupling, Figure 1.26, results in zero-field splitting (ZFS) of state degeneracies in addition to the ligand field effects on orbital degeneracy, which are related to the axial and rhombic distortions to the coordination sphere of the central metal.

The MCD spectra of low spin Fe(II) complexes are dominated by Faraday $\mathcal{A}_1$ terms since $S = 0$ due to the d^6 configuration and the groundstate is therefore nondegenerate. In contrast, the ground and excited states of low spin Fe(III) complexes split into Kramers doublets within an applied magnetic field with $M_S = -1/2$ and $M_S = +1/2$, since $S = 1/2$ due to the d^5 configuration.

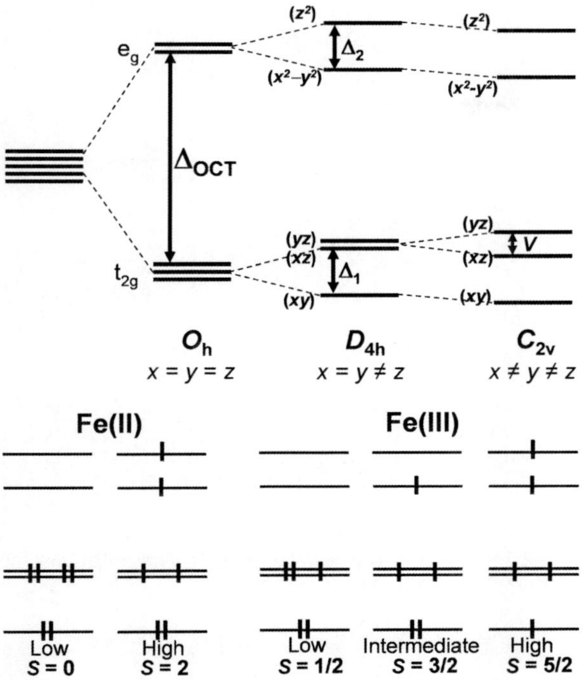

Figure 1.26 The crystal field splitting of Fe(II) and Fe(III) *d* orbitals in cubic O_h, tetragonal D_{4h} and rhombic C_{2v} environments (top). Only the latter two are typically observed for haem complexes. The axial and rhombic ligand field parameters are denoted as Δ and *V*. The low, intermediate and high spin states for Fe(II) and Fe(III) are based on a tetragonal environment (bottom).

At high temperature and low field, the lcp and rcp bands tend to cancel, resulting in a weak MCD signal, Figure 1.27. As the temperature is lowered or the applied field is increased, the Boltzmann population of the higher sublevel decreases and the $\mathcal{C}_0$ term, therefore, increases in intensity. During measurement of vitreous solutions at cryogenic temperatures (1–4 K) and high field strengths (4–7 T), the lowest component of the groundstate is often preferentially populated and the intensity of the MCD signal saturates. Although it is possible in principle to derive the groundstate g-factor value by measuring the $\mathcal{C}_0/\mathcal{D}_0$ ratio based on a method of moments analysis, this is usually not practical, since the absorption bands lack a baseline to baseline band envelope, or overlap significantly with bands associated with other metal centres or chromophores. It should be noted that, in some instances, excited state parameters also have to be carefully considered. For example, when groundstate degeneracy (and the associated Boltzmann distribution of population across microstates within an applied magnetic field) is introduced by spin degeneracy into what would otherwise be a Faraday $\mathcal{A}_1$ term, Figure 1.28, a temperature dependent, derivative-shaped signal is observed to have $\mathcal{C}_0$ term intensity.

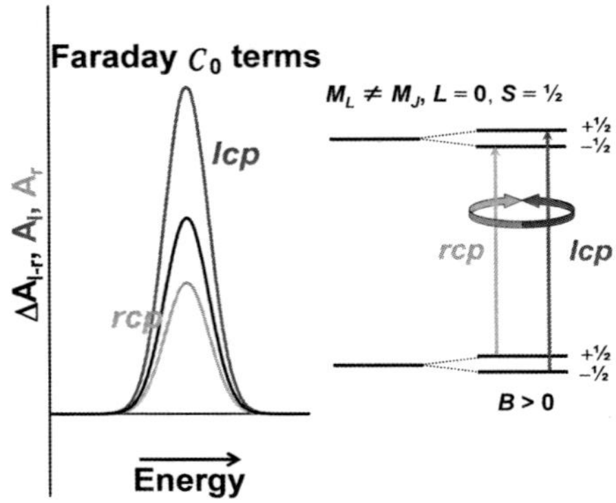

Figure 1.27 Selection rules (right) for the absorption of circularly polarised light based on a transition between an orbitally degenerate groundstate and an orbitally nondegenerate excited state within a magnetic field applied parallel to the axis of light propagation. A Gaussian-shaped Faraday C_0 term is observed due to the separation of the band centres for the absorption of lcp and rcp light. The C_0 term has a $1/kT$ temperature dependence since the groundstate microstates are populated according to a Boltzmann distribution.

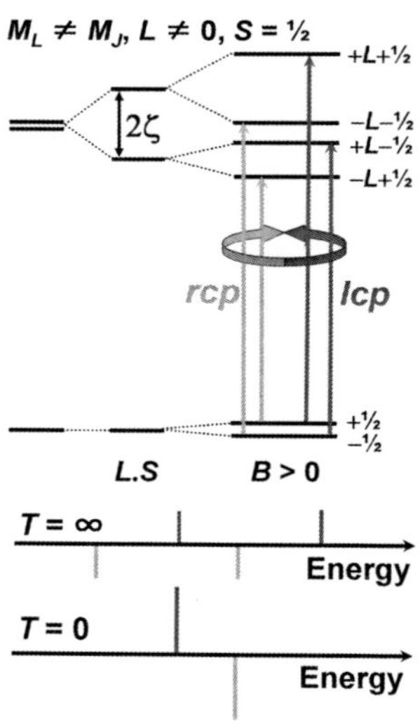

Figure 1.28 The origin of the Faraday C_0 term from Kramers doublets for the ground and excited states of a paramagnetic complex with $S = 1/2$, a non-orbitally degenerate groundstate and orbitally degenerate excited state (top). Temperature dependent, derivative-shaped signals can result when the groundstate is orbitally nondegenerate and the excited state is orbitally degenerate (bottom).

Schatz *et al.*[24] demonstrated that for a D_{4h} symmetry complex with the $S = 1/2$ Kramers doublet groundstates associated with low spin Fe(III) porphyrinoids, the zeroth moment of C_0 term intensity for an axial complex with $g_x \approx g_y$ in the linear limit is given by:

$$\langle \Delta\varepsilon \rangle_0 = \frac{\mathrm{K}\mu_\mathrm{B} B}{3\mathrm{k}T} \times \frac{g_{\parallel} \left[\boldsymbol{\mu}_{+}^{\mathrm{AJ}}\right]^2}{2} \tag{1.51}$$

where $\boldsymbol{\mu}_{+}^{\mathrm{AJ}} \equiv {}^{1}\!\Big/\!\sqrt{2(\boldsymbol{\mu}_x + i\boldsymbol{\mu}_y)}$. At saturation:

$$\langle \Delta\varepsilon \rangle_0 = \left(\mathrm{K}\left[\boldsymbol{\mu}_{+}^{\mathrm{AJ}}\right]^2\right) \frac{\cosh(\alpha)}{\sinh^3(\alpha)} \times \left(\frac{\sinh(2\alpha)}{4} - \frac{\alpha}{2}\right) \tag{1.52}$$

where K is a proportionality factor and α is related to the degree of anisotropy in the g-factor values:

$$\frac{g_{\parallel}}{g} = \sigma = \cosh(\alpha) \tag{1.53}$$

$$\frac{1}{\mathrm{L}} = \frac{\cosh(\alpha)}{\sinh^3(\alpha)} \times \left(\frac{\sinh(2\alpha)}{4} - \frac{\alpha}{2}\right) \tag{1.54}$$

The ratio of the slope to the MCD intensity at saturation defined by an intercept value, I, can be used to derive the expression:

$$g_{\parallel} = \frac{3\mathrm{I}}{\mathrm{L}} \tag{1.55}$$

Thomson[25] subsequently demonstrated that it is possible to use magnetisation curves of $\Delta\varepsilon$ plotted against $\mu_\mathrm{B}B/2\mathrm{k}T$, Figure 1.29, based on C_0 term temperature dependence studies at cryogenic temperatures (typically 2–50 K)

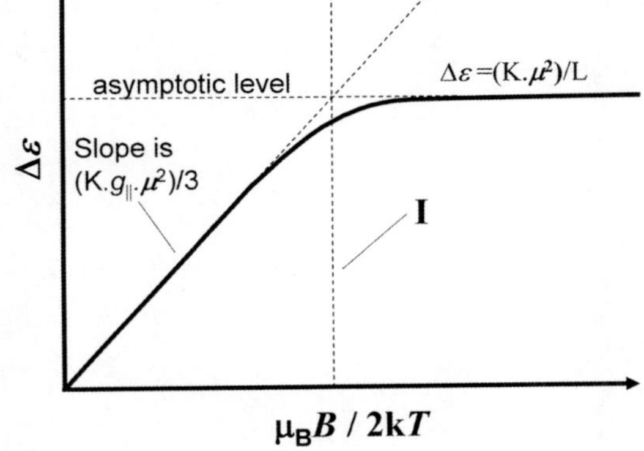

Figure 1.29 Analysis of an MCD magnetisation curve based on equations (1.51–1.55).

and applied field strengths of 1–6 T, to derive values for $g_{||}$ and $g_{\perp}$ based on the expressions:

$$g_{\text{av}} = \frac{1}{\sqrt{3}}\sqrt{g_x^2 + g_y^2 + g_z^2} \tag{1.56}$$

$$g\perp = \frac{1}{\sqrt{2}}\sqrt{g_x^2 + g_y^2} \tag{1.57}$$

In the case of the $S = 5/2$ groundstates of high spin Fe(III) complexes, there is a ZFS of the $M_S = \pm 1/2$, $\pm 3/2$, $\pm 5/2$ microstates due to spin–orbit coupling based on the axial (D) and rhombic (E) ZFS parameters within the spin Hamiltonian:

$$H = D\left(S_z^2 - \frac{1}{3}S^2\right) + E(S_x^2 - S_y^2) \tag{1.58}$$

of the central metal, Figure 1.30. In the early 1980s, Stephens and co-workers[26] demonstrated that the magnitude of D could be derived by fitting a plot of $\Delta\varepsilon$ against T^{-1} from cryogenic temperature MCD spectral data of a high spin Fe(III) tetraphenylporphyrin with a set of quadratic equations based on a least squares fitting approach, since the MCD signal is the sum of the MCD signals of the three sets of Kramers doublets which are populated according to a Boltzmann distribution based on the magnitude and sign of D. Study of the temperature dependence over a range where kT is comparable to the magnitude of the ZFS (typically 20 cm^{-1}) enables D to be calculated.

Analysis of $S = 2$ groundstates associated with high spin d^6 configurations of Fe(II) are considerably more challenging than the Kramers doublet based systems which arise from d^5 configurations, since the even spin groundstates are EPR silent. In 1980, Thomson and Johnson[25] reported multiple nested magnetisation curves, Figure 1.31, for the $S = 2$ groundstate of high spin Fe(II) myoglobin but did not attempt a detailed analysis due to the complexity of the field-induced mixing of the zero-field splitting components. During the 1990s, Solomon developed techniques for analysing variable temperature variable

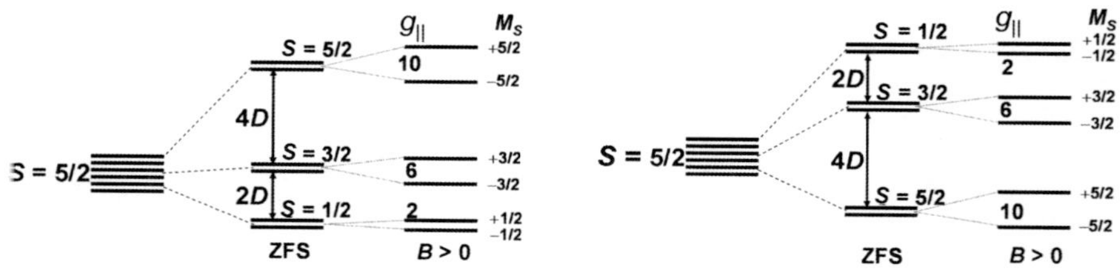

Figure 1.30 The zero-field splitting of the groundstate of a D_{4h} symmetry metal porphyrinoid complex with $S = 5/2$ based on the axial parameter, D. When D is positive the $S = 1/2$ microstates are stabilised (left), while if D is negative the $S = 5/2$ microstates are stabilised (right).

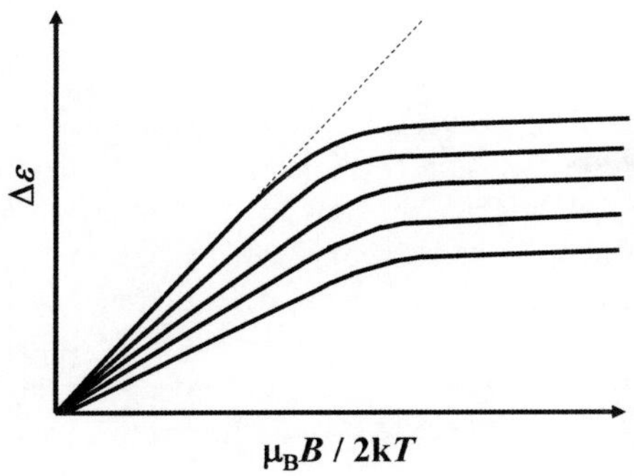

Figure 1.31 When there is a non-Kramers doublet $S = 2$ groundstate for high spin d^6 Fe(II) complexes, complex nested magnetisation curves are observed at different temperatures and field strengths based on equations (1.51–1.55) since $g_x \neq g_y \neq g_z$ due to the zero-field splitting of the groundstate.

field (VTVF) MCD spectra of non-Kramers doublet groundstates as part of a wide-ranging study of non-haem iron enzymes.[27] Solomon has demonstrated that it is possible to calculate values for the ligand field splitting properties, Δ and V, see Figure 1.26, of the t_{2g} set of orbitals of high spin $S = 2$ Fe(II) complexes from an analysis of nested magnetisation curves, Figure 1.31, based on experimentally obtained values of the zero-field rhombic splitting of the $M_S = \pm 2$ level.

Although the focus of this section has primarily been on iron complexes, since it is the most important transition metal in biochemistry and has therefore been studied most extensively by MCD spectroscopists, the techniques described can be readily applied to other transition metals.

1.3.5 Qualitative Analysis of MCD Spectra

It is important to bear in mind that a considerable amount of information can often be derived from MCD spectra even in the absence of temperature dependence studies and a detailed quantitative analysis of the Faraday terms. As will be described in greater detail in Chapter 9, the presence of derivative-shaped Faraday $\mathcal{A}_1$ in the spectra of high symmetry molecules and complexes can be used to identify the band centres of the main electronic bands within envelopes of extensively overlapping spectral bands. This can prove useful when validating the results of TD-DFT calculations, since it is often difficult to distinguish between bands which are electronic and vibrational in origin in the context of the electronic absorption spectrum. Although the analysis of Faraday $\mathcal{B}_0$ terms is usually less useful in this regard, Chapter 10 will describe how key information about the electronic structure can still be derived from sign sequences observed in MCD spectral data of aromatic and heteroaromatic π-systems when molecular orbital theory is used to describe the electronic structure and optical spectroscopy.

References

1. P. J. Stephens, *J. Phys. Chem.*, 1965, **43**, 4444.
2. S. B. Piepho and P. N. Schatz, *Group Theory in Spectroscopy with Applications to Magnetic Circular Dichroism*, Wiley, New York, 1983.
3. (a) J. L. Hughes, R. J. Pace and E. Krausz, *Chem. Phys. Lett.*, 2004, **385**, 116; (b) J. L. Hughes, R. Razeghifard, M. Logue, A. Oakley, T. Wydrzynski and E. Krausz, *J. Am. Chem. Soc.*, 2006, **128**, 3649.
4. (a) A. S. Davydov, *Zhur. Eksptl. I Teoret. Fiz.*, 1948, **18**, 210; (b) A. S. Davydov, *Theory of Molecular Excitons*, McGraw-Hill, New York, 1962.
5. M. Kasha, H. R. Rawls and M. A. El-Bayoumi, *Pure Appl. Chem.*, 1965, **11**, 371.
6. E. Charney, *The Molecular Basis of Optical Activity. Optical Rotatory Dispersion and Circular Dirchroism*, John Wiley & Sons, New York, 1979.
7. J. D. Watson and F. H. C. Crick, *Nature*, 1953, **171**, 737.
8. L. D. Bari, G. Pescitelli and P. Salvadori, *J. Am. Chem. Soc.*, 1999, **121**, 7998.
9. I. Tinoco, *Adv. Chem. Phys.*, 1962, **4**, 113.
10. P. E. Shipper, *J. Am. Chem. Soc.*, 1979, **101**, 6826.
11. N. Kobayashi, R. Higashi, B. C. Titeca, F. Lamote and A. Ceulemans, *J. Am. Chem. Soc.*, 1999, **121**, 12018.
12. (a) V. E. Shashoua, *J. Am. Chem. Soc.*, 1965, **87**, 4044; (b) M. V. Volkenstein, J. A. Sharanov and A. K. Shemelin, *Nature*, 1966, **209**, 709; (c) V. E. Shashoua, *J. Am. Chem. Soc.*, 1964, **86**, 2109; (d) B. Briat, *Compt. Rend.*, 1964, **259**, 2408; (e) P. J. Stephens, *Inorg. Chem.*, 1965, **4**, 1690.
13. P. J. Stephens, W. Süetaka and P. N. Schatz, *J. Chem. Phys.*, 1966, **44**, 4592.
14. *Nobel Lectures, Physics 1901–1921*, Elsevier Publishing Company, Amsterdam, 1967.
15. (a) P. J. Stephens, *Adv. Chem. Phys.*, 1976, **35**, 197; (b) A. D. Buckingham and P. J. Stephens, *Ann. Rev. Phys. Chem.*, 1966, **17**, 399.
16. R. Serber, *Phys. Rev.*, 1932, **41**, 489.
17. C. H. Henry, S. E. Schatterly and C. P. Slichter, *Phys. Rev.*, 1965, **A137**, 583.
18. J. Michl, *Tetrahedron*, 1984, **40**, 3845.
19. B. Briat, D. A. Schooley, R. Records, E. Bunnenberg and C. Djerassi, *J. Am. Chem. Soc.*, 1967, **89**, 6170.
20. (a) P. N. Schatz, A. J. McCaffery, W. Süetaka, G. N. Henning, A. B. Ritchie and P. J. Stephens, *J. Chem. Phys.*, 1966, **45**, 722; (b) P. N. Schatz and A. J. McCaffery, *Quart. Rev. Chem. Soc.*, 1969, **23**, 552.
21. (a) P. J. Stephens, *J. Chem. Phys.*, 1970, **52**, 3489; (b) P. J. Stephens, *Ann. Rev. Phys. Chem.*, 1974, **25**, 201.
22. B. Briat, D. A. Schooley, R. Records, E. Bunnenberg and C. Djerassi, *J. Am. Chem. Soc.*, 1967, **89**, 7062.
23. (a) J. H. Dawson and D. M. Dooley in *Iron Porphyrins*, Vol. 4, ed., A. B. P. Lever and H.B. Gray, VCH, London, 1989, p. 1; (b) J. Cheek and J. Dawson in *The Porphyrin Handbook*, Vol. 7, ed. K. Kadish, K. Smith,

and R. Guilard, Academic Press, New York, 2000, p. 339; (c) M. L. Kirk and K. Peariso, *Curr. Opin. Chem. Biol.*, 2003, **7**, 20; (d) E. I. Solomon, E. G. Pavel, K. E. Loeb and C. Campochiaro, *Coord. Chem. Rev.*, 1995, **144**, 369.

24. P. N. Schatz, R. L. Mowery and E. R. Krausz, *Mol. Phys.*, 1978, **35**, 1537.
25. A. J. Thomson and M. K. Johnson, *Biochem. J.*, 1980, **191**, 411.
26. W. R. Browett, A. F. Fucaloro, T. V. Morgan and P. J. Stephens, *J. Am. Chem. Soc.*, 1983, **105**, 1868.
27. E. I. Solomon, T. C. Brunold, M. I. Davis, J. N. Kemsley, S.-K. Lee, N. Lehnert, F. Neese, A. J. Skulan, Y.-S. Yang and J. Zhou, *Chem. Rev.*, 2000, **100**, 235.
28. A. Tanatani, A. Yokoyama, I. Azumaya, Y. Takakura, C. Mitsui, M. Shiro, M. Uchiyama, A. Muranaka, N. Kobayashi and T. Yokozawa, *J. Am. Chem. Soc.*, 2005, **127**, 8553.

CHAPTER 2

Empirical Rules in CD Spectra and Absolute Configuration of Molecules

In this chapter, a few relatively reliable empirical rules are introduced that can be applied to the circular dichroism (CD) spectra of certain symmetrical chromophores based on the observed sign of the CD signal. In many cases the absolute configuration of a molecule can be inferred if the conformation is known, based solely on the application of an empirical rule, while in other cases the conformation can be predicted if the absolute configuration is known. The octant rule for ketones is perhaps the best known of these empirical rules and is based on an analysis of the sign of the R band in the 290–300 nm region, which is associated with an electric dipole forbidden but magnetic dipole allowed n → π* transition. Rules which have been developed to predict the sign of the CD bands in the 230–270 nm region of the spectra of substituted benzene compounds are also described.[1] These bands are associated with the electric and magnetic dipole forbidden ^{1}Lb band.

2.1 The Octant Rule

The octant rule can be used to determine the absolute configuration of a saturated ketone or aldehyde if its confirmation is known, or *vice versa*. Cyclohexanone provides a good example, see Figure 2.1. The three bonds of C1 are used to define the *yz*-plane and the origin of the three axes is placed at the centre of the C=O bond. An n → π transition in the 290–300 nm region results in a rotation of charge, since an electron moves from an oxygen nonbonding orbital (p_y) to a π* antibonding orbital formed through a linear combination of the oxygen and carbon p_x orbitals. This leads to a large induced magnetic dipole transition moment (mdtm), ***m***, along the C=O bond (*z*-axis), which

Circular Dichroism and Magnetic Circular Dichroism Spectroscopy for Organic Chemists
By Nagao Kobayashi, Atsuya Muranaka and John Mack

Published by the Royal Society of Chemistry, www.rsc.org

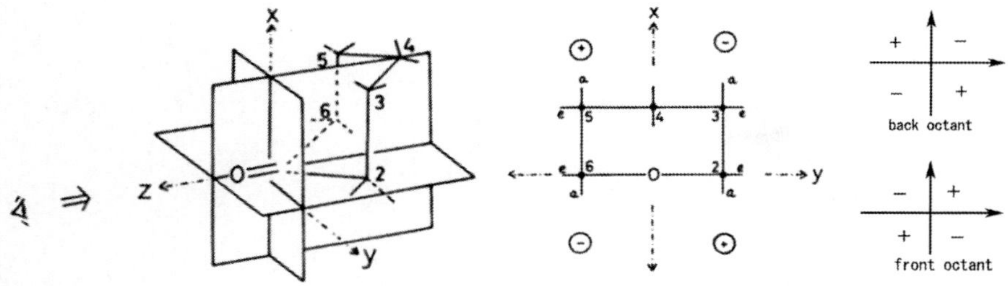

Figure 2.1 Octant rule diagram (left) and octant projection diagram (centre) applied to cyclohexanone. The C=O group is aligned along the intersection of the *xz*- and *yz*-planes and is viewed from the O atom towards the C atom. The sign of the contributions to the positions of the cyclohexanone carbons to the rotatory strength of the n → π* transition can be determined by plotting their positions onto a back octant projection diagram (right). It should be noted that the signs of the front octants are the opposite of those of the back octants. [Reprinted with permission from *J. Chem. Educ.*, 1975, **52**, 774. Copyright 1975, American Chemical Society.]

couples with a weak electric dipole transition moment (edtm), $\boldsymbol{\mu}$, whose intensity is borrowed from higher energy electric dipole-allowed π → π* or σ → σ* transitions by vibronic coupling, to produce a non-zero rotational strength. When this molecule is viewed from the carbonyl oxygen atom along the *z*-axis, the saturated cyclohexanone ring and substituents can be divided into eight octants based on the *xy*-, *xz*- and *yz*-planes. The perspective along the carbonyl bond is used to define the areas which lie behind or in front of the *xy*-plane as the back and front octants, respectively, see Figure 2.1.

Cyclohexanone possesses a mirror plane so it lacks a chiral centre, but this changes when substituents are introduced in an asymmetrical manner. These substituents are usually referred to as perturbers. The sign of the perturbation to the rotatory strength of the n → π* band, which is introduced by a substituent group, is determined by which octant it lies in. The contributions of different perturbers are assumed to be additive. Their relative magnitudes vary from substituent to substituent based primarily on their polarisability. Proximity to the nodal surfaces is also an important factor. Substituents which lie on the *xy*-, *xz*- and *yz*-planes do not contribute to the optical activity. In general, the octant contributions fall off with increasing distance from the carbonyl chromophore. The introduction of substituents at the rear upper left and lower right octants, Figure 2.1, leads to a positively signed band, while substituents located in the rear lower left and upper right octants result in a negatively signed band. In the context of the chair conformer of cyclohexanone, carbons 1, 2, 4 and 6 lie on octant planes and thus make no contribution. Carbons 3 and 5 lie in the upper left (+) and upper right (−) back octants equally displaced from the *xz*-symmetry plane, so they make no net contribution to the CD signal. When a methyl group is introduced to a β carbon to form (*3R*)-methylhexanone, the equatorial methyl group lies in the upper left (+) back octant and the net contribution to the CD signal is positive,[2] Figure 2.2.

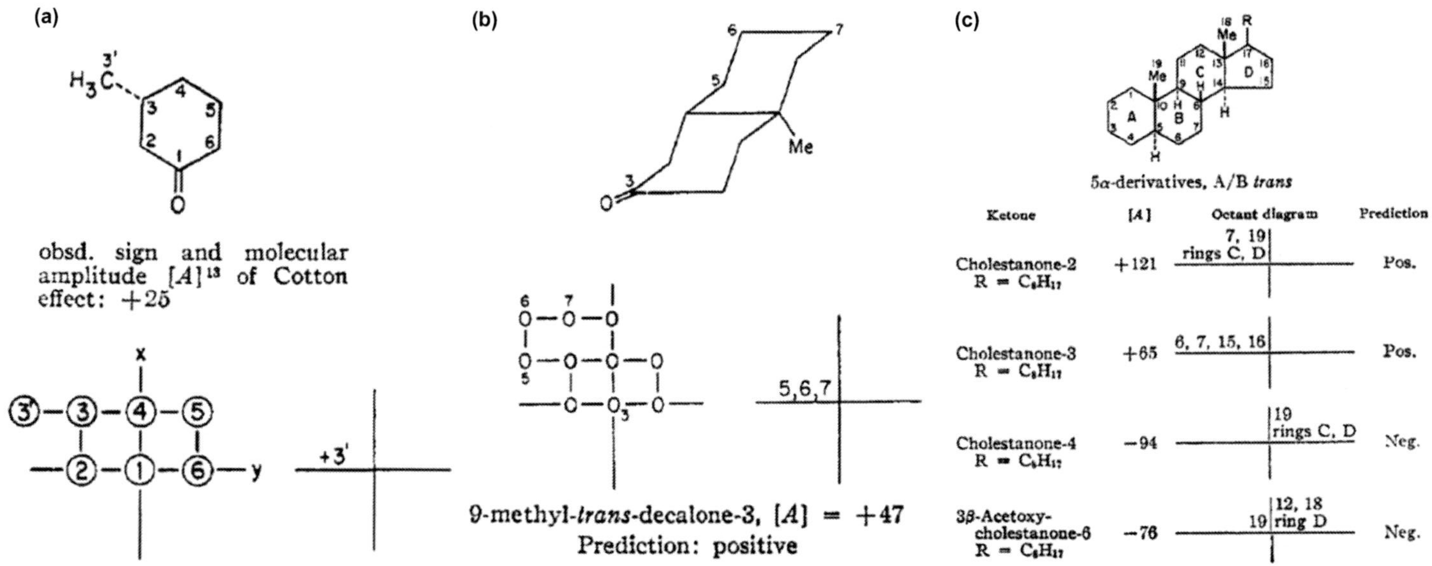

Figure 2.2 The application of the octant rule to (a) a substituted cyclohexanone, (b) a decalone and (c) a series of steroid structures. [Reprinted with permission from *J. Am. Chem. Soc.*, 1961, **83**, 4013. Copyright 1961, American Chemical Society.]

Considerably more complicated structures such as decalones and steroids were also analysed successfully using this approach when the octant rule was first formulated.

Unfortunately, despite these early successes, the application of the octant rule is complicated by the fact that the *xy*-plane bisecting the C=O bond was initially defined as being a planar surface for reasons of convenience rather than on any theoretical basis related to the symmetry of the MOs associated with the $n \rightarrow \pi^*$ transition.[3] Subsequent theoretical work has demonstrated that a concave surface is actually formed between the back and front octants,[3,4] Figure 2.3. In the early years of the application of the octant rule this flaw in the model led to the identification of several so called "antioctant" compounds. Even in the context of cyclohexanones, for example, alkyl substituents were sometimes found to result in anomalous signals. Kirk and Klyne[5] summarised the contribution of different types of alkyl group in this regard, Figure 2.4 and Table 2.1, and used the terms octant consignate (con.) and dissignate (dis.) to refer to whether the sign of the observed CD signal is or is not in accord with the octant rule prediction. The β-axial methyl group has a weak dissignate contribution of *ca.* 0.1, if it lies on a primary zig-zag of an alkyl group with only one or two C–C bonds, while similar dissignate signals have been reported for γ'-axial and γ'-equatorial groups, due to the concave nature of the third nodal surface,

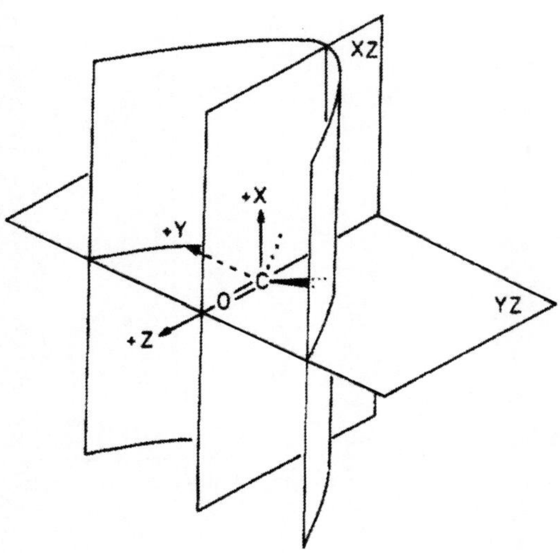

Figure 2.3 The revised octant rule[3] with the *xz*- and *yz*-octant planes determined by symmetry and thus unchanged relative to Figure 2.1. In contrast with Figure 2.1, the third nodal surface is defined theoretically as a concave surface when viewed from the O atom towards the C atom, which cuts in behind the C atom of the carbonyl and intersects the *yz*-plane approximately along the line $y = z + 2$. [Reprinted with permission from *J. Am. Chem. Soc.*, 1986, **108**, 4484. Copyright 1986, American Chemical Society.]

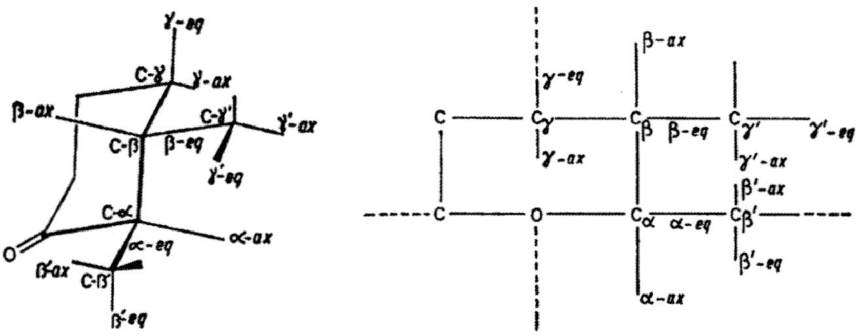

Figure 2.4 The application of the octant projection diagram to substituted cyclohexanones.[5] Vertical and horizontal dashed lines are used to highlight the *xz*- and *yz*-planes, and ax and eq refer, respectively, to axial and equatorial bonds. [Reprinted with permission from *J. Chem. Soc., Perkin Trans., 1*, 1974, 1076. Copyright 1975, Royal Society of Chemistry.]

Table 2.1 Contribution of alkyl groups to the Δε value of the R (n→π*) band in a range of different solvents.[5] [Reprinted with permission from *J. Chem. Soc., Perkin Trans. 1*, 1974, 1076. Copyright 1974, Royal Society of Chemistry.]

Methyl or alkyl group	*sign*	*hexane*	*dioxane*	*CH_3CN*	*MeOH (or EtOH)*
α-axial (>$CHCH_3$)	con.	1.2	1.3	1.4	1.5
α-axial(→CCH_3)	con.	1.6	1.65	1.65	1.7
α-equatorial	con.	0.2	0.2	0.2	0.15
β-axial (the 1st class)[a]	con.	0.45	0.5	0.55	0.6
β-axial (the 2nd class)[a]	dis.	0.1	0.1	0.1	0.1
β-equatorial	con.	0.5	0.55	0.55	0.6
γ-axial (in *trans*-2-decalone)[b]	—	0.25	0.25	0.2	0.2
γ-equatorial (in *cis*-2-decalone)[b]	—	0.1	0.1	0.1	0.1
β'-axial or β'-equatorial[c]	con.	0.2	0.2	0.2	0.2
γ'-axial	dis.	0.1	0.15	0.15	0.2
γ'-equatorial	dis.	0.6	0.7	0.8	0.9
β-equatorial-C_2H_5	con.	0.6	0.7	0.8	0.9
β-equatorial-*i*-C_3H_7	con.	0.75	0.85	1.0	1.1
β-equatorial-*t*-C_4H_9	con.	0.65	0.75	0.85	0.9

[a]When the β-axial methyl group lies on a zig-zag alkyl group beyond the first two C–C bonds (Cα–Cβ & Cβ–Cγ bonds), a consignate contribution of *ca.* 0.5 is expected (1st class), while on a zig-zag carbon chain consisting of one or two C–C bond(s), the β-axial group has a weak dissignate contribution of *ca.* 0.1 (2nd class).
[b]Contribution to reduce the |Δε| value in the decalone.
[c]Consignate contribution to the front octant.

Figure 2.3. It is noteworthy that the contribution of an α-equatorial $-CH_3$ group is non-zero, Table 2.1, and that it is also not safe to assume that a γ-equatorial or -axial group lies on the nodal plane.

When applying the octant rule, molecular models are often the best way to visualise the alignment of the octant planes. A Dreiding model is highly recommended in this regard. Although the octant rule is one of the more reliable empirical rules for determining the sign of the CD bands of organic molecules when two possible structures are predicted to result in CD signals of opposite sign, and it has wide applicability even for ketones consisting of five or six rings, it is important to review the known exceptions before any detailed analysis is attempted on this basis, since a more rigorous theoretical approach is often required.[3,4]

2.2 Benzene Sector and Benzene Chirality Rules

The electronic absorption spectra of benzene and many substituted benzene derivatives contain a weak ^1L_b band in the 230–270 nm region. In the context of benzene, the ^1L_b transition is electric and magnetic dipole forbidden and gains intensity through vibronic borrowing from allowed transitions at shorter wavelengths. When a substituent is introduced onto the periphery of the benzene ring, an edtm is induced, which lies normal to the subsituent attachment bond, and this in turn induces a mdtm perpendicular to the benzene ring. These induced ^1L_b transition moments can generate a CD signal based on dynamic and static coupling with chiral substituents.

2.2.1 Benzene Sector Rule[6]

In the absence of additional ring substituents, the sign of the ^1L_b band in the CD spectra of chiral benzene compounds is determined by the chiral centre immediately attached to the benzene ring. The sign of the ^1L_b band can, therefore, be readily predicted if the conformation of the chiral group at the attachment bond to the benzene ring is known. The empirical sector rule enables the summation of the contributions to the rotatory strength of the observed CD band that are made by each group attached to the chiral centre. In benzene compounds with a contiguous chiral centre containing a hydrogen atom as one of the substituents, the hydrogen atom eclipses the phenyl ring plane in the preferred conformation or nearly does so. Figure 2.5 contains quadrant projections, which provide the sign of the contribution by atoms or

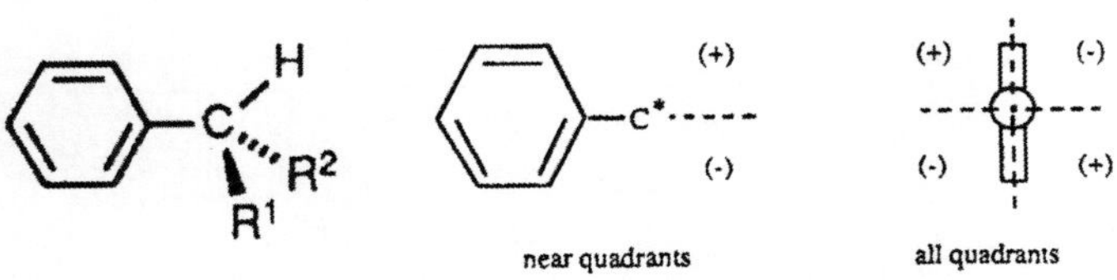

Figure 2.5 Octant projection diagram for benzene rings containing a single chiral substituent.[12] [Reprinted with permission from *J. Org. Chem.*, 1991, **56**, 432. Copyright 1987, American Chemical Society.]

groups attached to a chiral centre and lying in the near sectors in front of the benzene ring plane to the rotatory strength of the ^1L_b band in the CD spectrum of a mono-substituted benzene derivative. For atoms or groups in the far sectors behind the benzene ring plane, the sign of the vibronic contributions is reversed, while for those groups and atoms lying on sector boundaries, there is no contribution. Since the CD intensity results solely from vibronic borrowing in the context of mono-substituted benzene compounds, the sum of the contributions provides the sign of the observed CD signal for the ^1L_b band. The signs for the sectors were derived from the observed negative ^1L_b CD signals for (*R*)–α-phenylethyl alcohol and the assumption of a larger rotatory contribution for a methyl group, as compared to a hydroxyl group. The CD data for enantiopure compounds, in which one substituent at the contiguous chiral centre is a hydrogen atom and for which the absolute configurations are known, was used to derive a sequence for the summation of the rotatory contributions based on a ranking of the relative magnitudes of the rotatory contributions, $CO_2^- > (SH)$, $C(CH_3)_3 > CH_3 > COOH$, $COOCH_3 > NH_2$, N^+H_3, $N^+(CH_3)_3$, OH, OCH_3, CF_3, Cl, F. This sequence, when used in connection with the quadrant projections in Figure 2.5, can be used to establish the absolute configurations of chiral benzene compounds when one of the substituents at the contiguous chiral centre is a hydrogen atom. The effect of this ranking can be seen in Table 2.2.[7] For compound (*R*)–**1a**, the $-NH_2$ group is in the far (+) sector while the $-CH_3$ group lies in the near (−) sector. Since the ranking of the $-CH_3$ group is higher than that of the $-NH_2$ group, a negatively signed CD band is observed. Conversely in the context of (*S*)–**1a**, where the $-CH_3$ and $-NH_2$ groups are exchanged, a CD signal of opposite sign is expected. As anticipated, when the $-CH_3$ group is substituted by a $-C(CH_3)_3$ group, the CD intensity increases without changing the CD sign.

The application of the benzene sector rule to phenylcarbinamines and phenylcarbinols is relatively straightforward, since only the conformer in which the hydrogen atom at the chiral centre eclipses the benzene ring plane needs to be considered. When the rule is applied to benzylcarbinamines, benzylcarbinols and related compounds with a stereogenic centre separated from a benzene ring by a methylene group, Table 2.3, however, there is conformational mobility of the attached groups with respect to the benzene ring. The equilibrium between conformers (−)-**a** and (+)-**b** shown in Figure 2.6 must be considered when determining the sign of the rotatory strength of the ^1L_b band in the CD

Table 2.2 ^1L_b Band origin (0–0 band) maxima for phenylcarbinamines.

	R1	*R2*	*absorption [λ, nm (ε)]*	*CD [λ, nm (10^2Δε)]*	*solvent*	*ref.*
(*R*)–1a	NH_2	CH_3	267(86)	268(−11)	methanol	7a
(*S*)–1a	CH_3	NH_2	267(86)	268(+11)	methanol	7a
(*S*)–1b	CH_3	NH_3Cl	267(120)	267(+8.2)	methanol	7a
(*S*)–2a	$C(CH_3)_3$	NH_2	269(50)	268(+34)	isooctane	7b
(*S*)–2b	$C(CH_3)_3$	NH_3Cl	266(180)	266(+3.6)	hexafluoro-2-propanol	7b

Table 2.3 ^1L_b 0–0 Band maxima for some benzylcarbinamines, benzylcarbine-amine salts and benzylcarbinols.[6a] The relative alignments of R_1 and R_2 are shown in Figure 2.6. [Reprinted with permission from *Chem. Rev.* 1998, **98**, 1709. Copyright 1998, American Chemical Society.]

Config	R_1	R_2	*solvent*	*absorption [λ,* nm *(ε)^a]*	*CD [λ,* nm *(10^{2}Δε)^b]*
(R)–	CH_3	NH_2	0.1 M KOH/MeOH	267(120)	269(–3.0)
(R)–	CH_3	N^+H_3 (HCl salt)	MeOH	266(77)	267(–3.3)
(R)–	CH_3	OH	MeOH	267(120)	268(–5.5)
(S)–	C_5H_{11}	NH_2	cyclohexane		269(+4.5)
(R)–	CH_3	$NHCH_3$	95% EtOH	268(170)	268(–9.2)
(R)–	CH_3	$N^+H_2CH_3$ (HCl salt)	95% EtOH	268(330)	268(–7.1)
(R)–	CH_3	$N(CH_3)_2$	95% EtOH	269(150)	269(–10)
(R)–	CH_3	$N^+H(CH_3)_2$ (HCl salt)	95% EtOH	268(81)	267(–9.1)
(R)–	CH_3	$N^+(CH_3)_3$ (HCl salt)	95% EtOH	268(87)	273(–13)

[a]Molar absorptivity.
[b]Molar dichroic absorption.

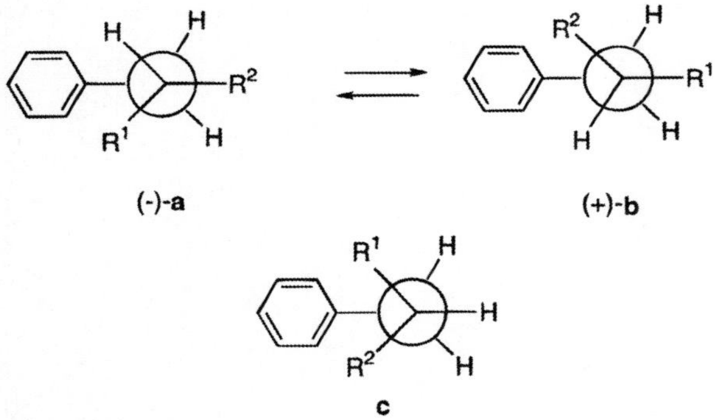

Figure 2.6 Only conformers (−)-**a** and (+)-**b** need to be considered when applying the benzene sector rule to benzylcarbinamines and benzylcarbinols, since the gauche interactions of conformer **c** lead to a higher energy. [Reprinted with permission from *J. Am. Chem. Soc.*, 1996, **118**, 7694. Copyright 1987, American Chemical Society.]

spectrum to the absolute configuration of such compounds. In chiral benzylcarbinamines and benzylcarbinols, the two lowest-energy conformations are those in which the C–C bond of the substituent is orthogonal to the benzene ring plane. Conformer **c** need not be considered, Figure 2.6, since the gauche interactions of the R_1 and R_2 groups with the benzene ring result in a significantly higher energy structure. Since the contribution of the hydrogen atom attached at the chiral centre is not significant,[8] and that of a group lying on a sector boundary is also small, the near sector rule that the upper and lower sectors result in positive and negative rotatory contributions, respectively, can be used to determine the sign of the rotatory strength of the ^1L_b band in the CD spectrum. Since the methyl group lies in the near lower (−) sector in most cases n Table 2.3, while the –OH, amino and protonated amino groups lie on sector boundaries, the CD signals would be predicted to be negatively signed. The experimental data, therefore, provide strong evidence that the compounds adopt the (−)-**a** conformation, Figure 2.6.

2.2.2 Benzene Chirality Rule

Phenylethylamine is optically active owing to the presence of the chiral side chain. When a second substituent group is introduced to the benzene ring, the CD signal can change in both intensity and sign, owing to the presence of an induced edtm normal to the attachment bond of the substituent. The magnitude of the induced edtm is related to the spectroscopic moment[9] of the substituent and can be resolved into component bond transition moments,[10,11] Figure 2.7.[12] The bond moments induced by the chiral group (solid lines) are single-handed, while those induced by the additional ring substituent can vary, depending on the position of substitution (*ortho*, *meta* and *para*) and the presence of a positive or negative partial charge on the substituents. An example is shown in Figure 2.7. Prior to the introduction of a Cl atom, the CD spectrum of

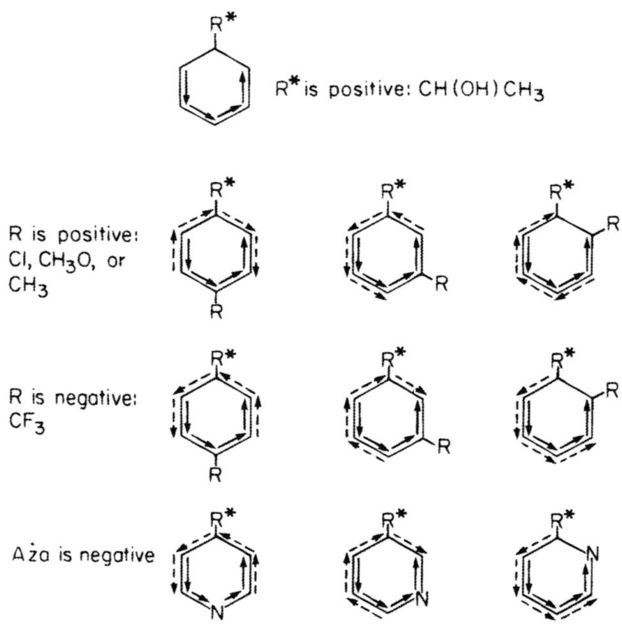

Figure 2.7 The induced bond moment directions for chiral benzene compounds substituted with groups having positive and negative spectroscopic moments. Solid arrows represent the bond moments induced by a chiral group, while the dashed arrows represent those induced by an additional ring substituent.[12] [Reprinted with permission from *J. Org. Chem.*, 1987, **52**, 3386. Copyright 1987, American Chemical Society.]

(*R*)–α-phenylethyl alcohol contains a negatively signed ^1L_b band, as would be anticipated based on the benzene sector rule. When the Cl atom is introduced at the *ortho* or *para* positions, resulting in a positive spectroscopic moment, the alignment of the induced bond transition moments opposes that of the chiral group, Figure 2.7, thus reducing the intensity of the negatively signed ^1L_b band. In contrast, when the Cl atom is introduced at the *meta* position there is an intensification of the negatively signed band, since the induced bond transition moments of the chiral group and the Cl atom are aligned with each other. As depicted in Figure 2.7, the introduction of a nitrogen heteroatom into the π-system or of a –CN or $-CF_3$ substituent group with a negative spectroscopic moment leads to the reverse effect. When the vibronic and induced contributions to the CD intensity of the ^1L_b bands have the same alignment, Figure 2.7, the sign of a particular ring-substituted benzene compound can be predicted with certainty. When these contributions have opposite alignments, the prediction of the sign can be somewhat problematic.[13] It should be noted, however, that the CD spectra of all of the phenylalkylcarbinamines and carbinols that have been reported to date with substituents at the *ortho*- and *para*-positions which have a positive spectroscopic moment (e.g. $-CH_3$, –Cl, –Br, –OH, $-OCH_3$), contain ^1L_b bands of opposite sign to that of the unsubstituted parent compound, Figure 2.8. In contrast, the sign of the ^1L_b

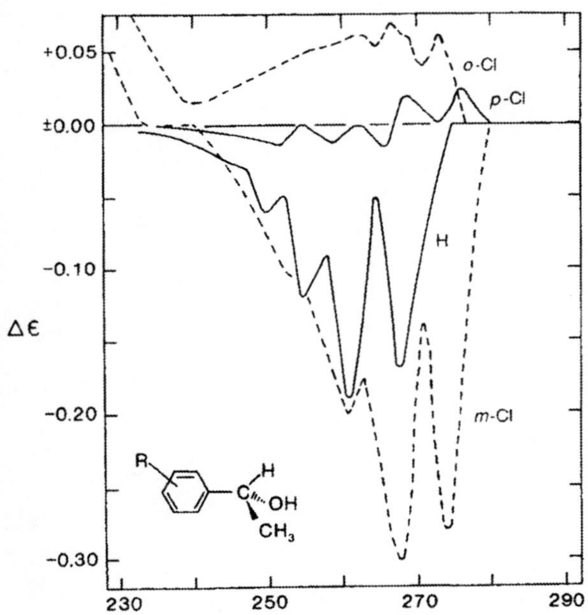

Figure 2.8 The CD spectra of (*R*)-α-phenylethylalcohol, (*R*)-α-(*p*-chlorophenyl)-ethylalcohol, (*R*)-α-(*m*-chlorophenyl)ethylalcohol and (*R*)-α-(*o*-chlorophenyl)ethylalcohol in methanol.[6] [Reprinted with permission from *J. Am. Chem. Soc.*, 1990, **112**, 5741. Copyright 1990, American Chemical Society.]

band remains unchanged in the case of phenylalkylcarbinamines and carbinols that are substituted by groups with a negative spectroscopic moment, such as CF_3 and CN, at the *meta* position.

References

1. B. Nordén, A. Rodger and T. Dafforn, *Linear Dichroism & Circular Dichroism: A Texbook of Polarized-Light Spectroscopy*, Royal Society of Chemistry, London, 2010.
2. W. Moffitt, A. Woodward, A. Moscowitz, W. Klyne and C. Djerassi, *J. Am. Chem. Soc.*, 1961, **83**, 4013.
3. T. D. Bouman and D. A. Lightner, *J. Am. Chem. Soc.*, 1976, **98**, 3145.
4. D. A. Lightner, in *Circular Dichroism*, ed. N. Berova, K. Nakanishi and R. W. Woody, Wiley-VCH, New York, 2000, p. 261.
5. D. N. Kirk and W. Klyne, *J. Chem. Soc., Perkin Trans. 1*, 1974, 1076.
6. (a) H. E. Smith, *Chem. Rev.*, 1998, **98**, 1709; (b) H. E. Smith, in *Circular Dichroism*, ed. N. Berova, K. Nakanishi and R. W. Woody, Wiley-VCH, New York, 2000, p. 397.
7. (a) H. E. Smith, J. R. Neergaard, T. de Pauli and F.-M. Chen, *J. Am. Chem. Soc.*, 1983, **105**, 1578; (b) W. C. Johnson, Jr., L. P. Fontana and H. E. Smith, *J. Am. Chem. Soc.*, 1987, **109**, 3361.
8. W. H. Inskeep, D. W. Miles and H. Eyring, *J. Am. Chem. Soc.*, 1970, **92**, 3866.
9. (a) J. R. Platt, *J. Chem. Phys.*, 1951, **19**, 263; (b) J. Petruska, *J. Chem. Phys.*, 1961, **34**, 1120.
10. S. F. Mason, *J. Chem. Soc., Chem. Commun.*, 1973, **239**.
11. J. Sagiv, *Tetrahedron*, 1977, **33**, 2315.
12. L. P. Fontana and H. Smith, *J. Org. Chem.*, 1987, **52**, 3386.
13. S. T. Pickard and H. E. Smith, *J. Am. Chem. Soc.*, 1996, **118**, 7694.

CHAPTER 3

Representative Systems Analysed by the Exciton Coupling Method

The circular dichroism (CD) bands which arise from exciton couplets have been widely studied in order to determine the configuration and conformation of natural products and organic compounds. When two close-lying chromophores are arranged obliquely in space, two absorption bands are normally anticipated at longer and shorter wavelengths than the absorption peak of the monomer. It should be noted, however, that these two peaks cannot always be clearly identified. From the standpoint of theoretical analysis, the use of two identical chromophores in a C_2 or lower symmetry arrangement is particularly important in the context of CD spectroscopy. Although they are not necessarily always stated explicitly, there are several points that should be taken into consideration when using this approach:

1. The chromophores must retain independence with no significant orbital overlap. When the chromophores are mutually conjugated, the analysis often turns out to be incorrect.
2. Ideally there should be large absorption coefficients in a wavelength region where the substrate does not absorb significantly.
3. The alignment of the edtms of the chromophores should be well established.
4. The chirality of the exciton in space should be determined unambiguously, including both the configuration and the conformation.

Several examples of the use of exciton coupling theory will now be described in detail.

Circular Dichroism and Magnetic Circular Dichroism Spectroscopy for Organic Chemists
By Nagao Kobayashi, Atsuya Muranaka and John Mack

Published by the Royal Society of Chemistry, www.rsc.org

3.1 Exciton Coupling among Fewer than Four Identical Chromophores

3.1.1 Spectra of 5α-Cholestane-2β,3β- and 3β,6β-diol *Bis*(*p*-dimethylamino-benzoate)[1,2]

p-Dimethylbenzoic acid can be used to esterify the diol groups in the structures of 5α-cholestane-2β,3β-diols so that chromophores are introduced, which can interact to form exciton couplets. This represents one of the best systems for describing the theory of exciton coupling in the context of CD spectroscopy. The lowest energy band (*ca.* 310 nm) with a large ε value of *ca.* 30 000–31 000 has an edtm polarised along the long axis of the molecule (*i.e.* along the N atom of the dimethylamino group and the C atom of the carboxyl group). The electronic absorption and CD spectra exhibit the typical characteristics of an exciton coupled system,[1] Figure 3.1. As expected, a derivative-shaped CD signal is observed for the longest wavelength ^1L_a absorption band (intramolecular $\pi \rightarrow \pi^*$ charge-transfer band). The ratio of the integrated intensity of the lower and higher energy CD bands is approximately 1. The Δε values are very large (generally more than one order of magnitude larger than those of a system containing only one dimethylaminobenzoate) and the Δε ratio of the two CD bands is *ca.* −2. The two OH groups were determined to adopt a right-handed helicity (i.e. a clockwise twist), since the sign sequence of the CD bands is positive to negative when viewed from longer wavelength. Figure 3.2 contains the spectra of 5α-cholestane-3β,6β-diol *bis*(*p*-dimethylaminobenzoate).[2] In this compound, the two dimethylaminobenzo groups are linked at the 3β and 6β positions of the 5α-cholestanediol, rather than at the 2β and 3β positions as is the case in Figure 3.1. Since the CD band intensity associated with exciton coupling is inversely proportional to the square of the distance between the two chromophores, the signal observed for 5α-cholestane-3β,6β-diol *bis*(*p*-dimethylaminobenzoate) is much weaker than that observed in Figure 3.1. The minus-to-plus sign sequence that is observed in ascending energy terms in the CD intensity is the inverse of that observed for the compound in Figure 3.1. This indicates that the two OH groups at the 3β and 6β positions form a left-handed helicity (*i.e.* an anticlockwise twist). The Δε ratio of the lower and higher energy CD bands is *ca.* −2 as in Figure 3.1.

Figures 3.1 and 3.2 demonstrate that the CD curves can be reproduced by adding absorption envelopes of opposite sign separated by a Davydov split of $2V_{ij}$, see equation (1.9).[1] The sign and intensity of an exciton couplet signal depends on both the distance between the two chromophores and the angle that they form. According to calculations by Harada *et al.*,[1] when the mutual distance is fixed, the absolute value changes, as shown in Figure 3.3, and a minimum or maximum in intensity is observed when the angle is close to 70°.

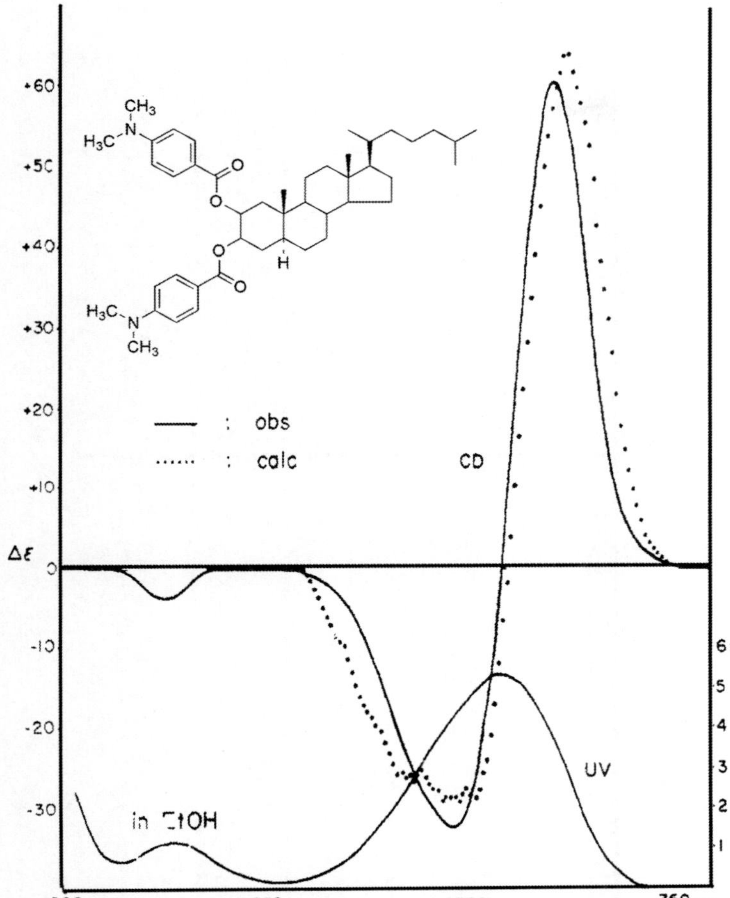

Figure 3.1 Electronic absorption and calculated and observed CD spectra of 5α-cholestane-2β,3β-diol *bis*-(*p*-dimethylaminobenzoate).[1] [Reprinted with permission from *J. Am. Chem. Soc.*, 1975, **97**, 5345. Copyright 1975, American Chemical Society.]

3.1.2 Spectra of (6*R*,15*R*)-(+)-6,15-Dihydro-6,15-ethanonaphtho[2,3–*c*]pentaphene

(6*R*,15*R*)-(+)-6,15-dihydro-6,15-ethanonaphtho[2,3–*c*]pentaphene contains two anthracene moieties in a rigid C_2 symmetry arrangement, Figure 3.4. The electronic absorption spectrum of anthracene contains a very strong (ε = *ca.* 204 000) long axis-polarised ^1B_b band in the 250–260 nm region and a weak (ε = *ca.* 11 000) short axis-polarised ^1L_a band between 300 and 400 nm. As a

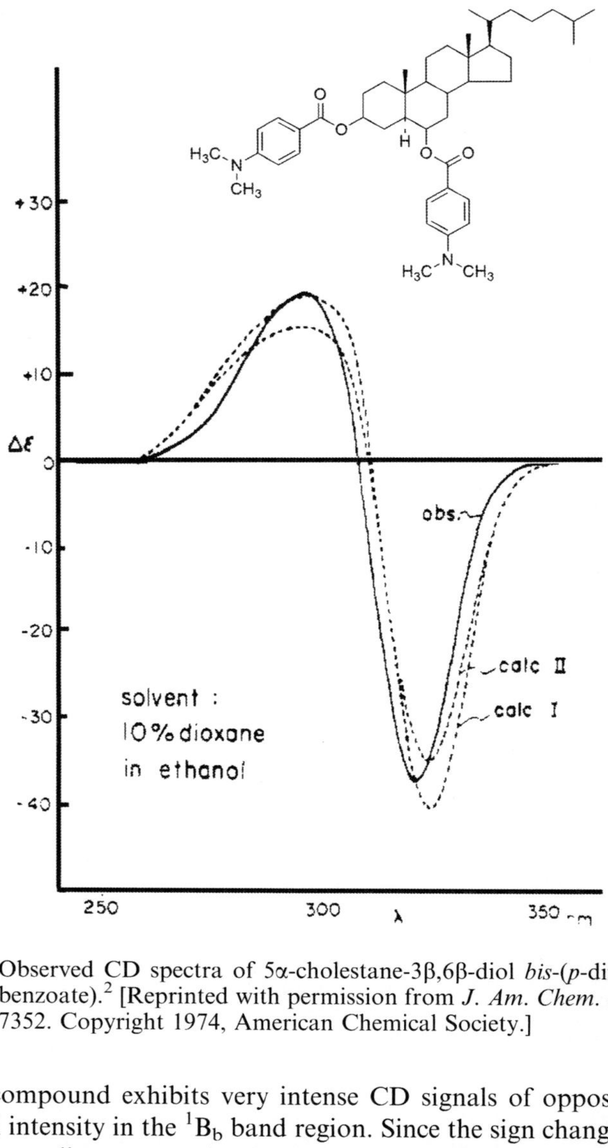

Figure 3.2 Observed CD spectra of 5α-cholestane-3β,6β-diol *bis*-(*p*-dimethylamino-benzoate).[2] [Reprinted with permission from *J. Am. Chem. Soc.* 1974, **96**, 7352. Copyright 1974, American Chemical Society.]

result, this compound exhibits very intense CD signals of opposite sign and almost equal intensity in the ^1B_b band region. Since the sign changes from plus to minus in ascending energy, the absolute configuration of this compound was determined to be *(6R,15R)*.[3]

It should be noted that in Figure 3.4 the inflection point of the CD curve does not match the absorption maximum. This can be rationalised by considering point dipole moments for the long axis-polarised ^1B_b transition placed at the centre of each anthracene moiety. As shown in Figure 3.5, in the original mode of coupling, ^1B_b transition moments are anti-parallel to each other and the electronic interaction energy estimated by the point dipole approximation, equations (1.9) and (1.10), has a value that is definitely positive. The two edtms

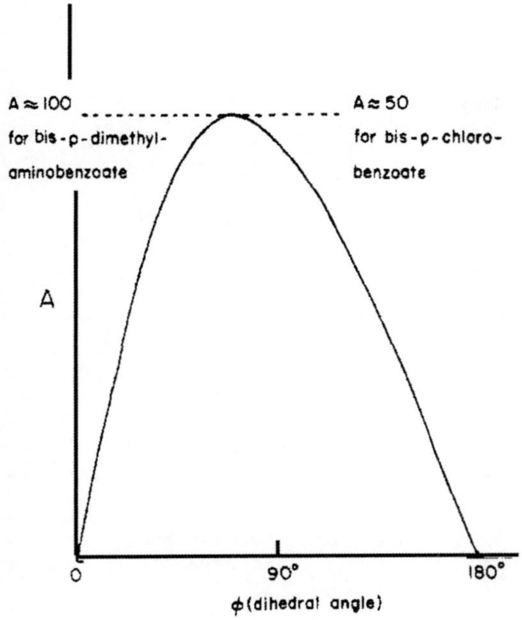

Figure 3.3 Relationship between dihedral angle and CD amplitude $A(\Delta\varepsilon_1 - \Delta\varepsilon_2)$ of vicinal *bis*(*para*-substituted benzoate).[1] [Reprinted with permission from *J. Am. Chem. Soc.*, 1975, **97**, 5345. Copyright 1975, American Chemical Society.]

of the α state at longer wavelength are in phase, while those of the β state at shorter wavelength are out of phase. Since the angle between the two vectors is determined by X-ray analysis to be 151°, the in-phase combination of edtms results in a larger net moment value than that of the out-of-phase combination. The ratio of absorption intensity (dipole strength) was, therefore, calculated to be *ca.* 15 : 1 on the longer and shorter wavelength sides, Figure 3.5. Hence, most of the absorption intensity is shifted to the position of the lower energy CD band, giving the impression that the inflection wavelength of the CD spectrum is located at shorter wavelength than the band centre of the absorption peak. This apparent discrepancy is, therefore, the pattern that would be anticipated for this system based on an exciton coupling type interaction, since the C_2 symmetry dictates that the transition moment of the β state (A symmetry) is polarised along the C_2 axis, while the transition moment of the α state (B symmetry) is polarised in a perpendicular direction.

3.1.3 Additivity Relationship in the Amplitudes of Exciton-Split CD Curves of Sugar Benzoates[4,5]

When more than two identical chromophores lie close together in a well-defined geometry, the amplitude ($A = \Delta\varepsilon_1 - \Delta\varepsilon_2$) of the peak and trough of the split CD

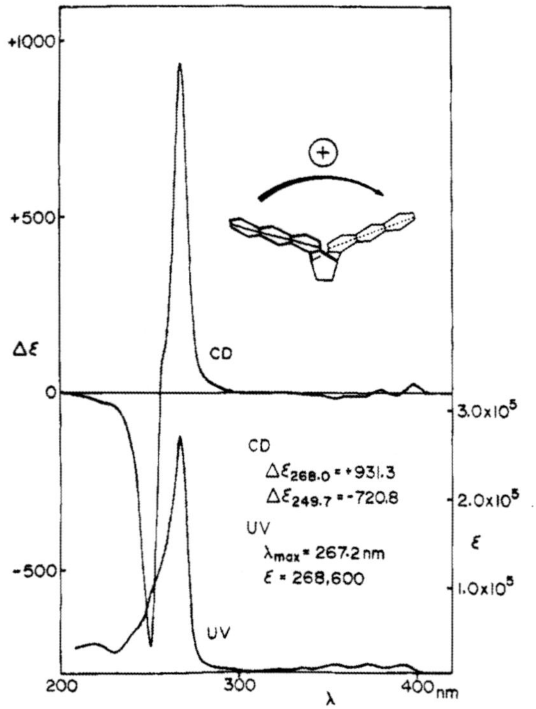

Figure 3.4 Electronic absorption and CD spectra of (6*R*,15*R*)-(+)-6,15-dihydro-6-15-ethanonaphtho[2,3*c*]pentaphene.[3a] [Reprinted with permission from *J. Am. Chem. Soc.*, 1976, **98**, 5408. Copyright 1976, American Chemical Society.]

signal can be approximated on the basis of an additivity relationship in which the known amplitudes for each pair of chromphores are added together to provide an estimate of the total observed signal. The validity of this approach has been demonstrated by systematic investigation. For example, the CD spectra of over 40 sugar benzoates have been studied. In the case of molecules containing three interacting chromophores, I, II and III, the *A* value can be approximated by the summation of the component *A* values: $A(\text{total}) = ca.\ A(\text{I/II}) + A(\text{II/III}) + A(\text{III/I})$. Figure 3.6 contains the *A* values of various pyranose *p*-bromobenzoates collected experimentally in methanol. The signs of the *A* values depend on the chiralities of the benzoate groups, which are indicated by solid circles. In this figure, the compounds are divided into six groups depending on the position of the benzoates:

Class I. The 1,2-ee and 1,2-ea dibenzoates (**1–3**) have dihedral angles of *ca.* 60°. The *A* values are 62.

Class II. The CD signal of the 1,2-aa dibenzoate (**4**) exhibits a small splitting based on 1,3-diaxial interactions with 1a–OMe, 4b–H and

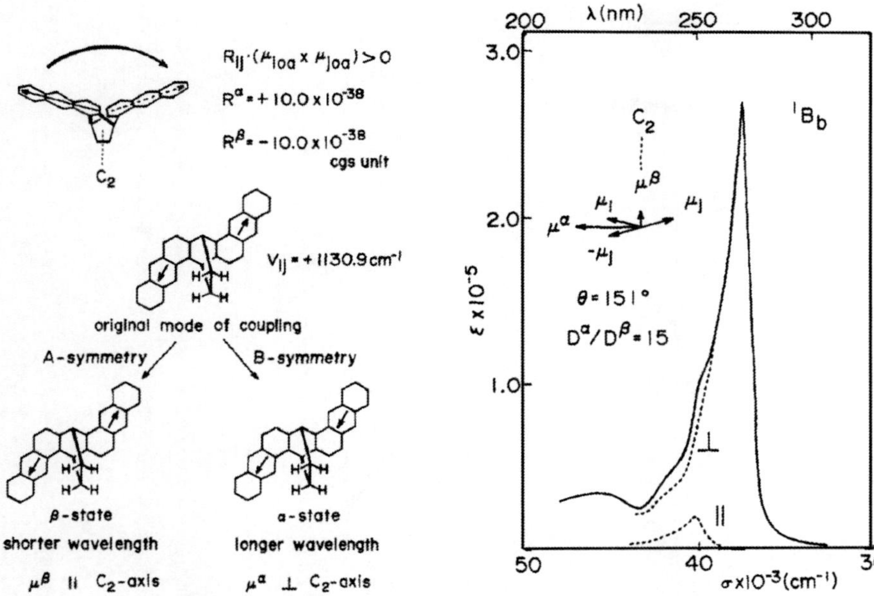

Figure 3.5 Rotational strength and coupling modes of two electronic transition moments for the ^1B_b transition of (6*R*,15*R*)-(+)-6,15-dihydro-6-15-ethanonaphtho[2,3-*c*]pentaphene (left). Calculated spectra of the ^1B_b transition of (6*R*,15*R*)-(+)-6,15-dihydro-6-15-ethanonaphtho[2,3*c*]pentaphene. The ⊥ and // symbols indicate the absorption perpendicular and parallel to the C_2 axis, respectively. The observed experimental spectrum is plotted with a solid line, while the calculated component ⊥ and // polarised signals are plotted with dashed lines. Since the angle between two moments is 151°, the ratio μ^{α}/μ^{β} is $\sqrt{15}$ and therefore $D^{\alpha}/D^{\beta}=15$.[3b] [Reprinted with permission from *J. Am. Chem. Soc.*, 1978, **100**, 4029. Copyright 1978, American Chemical Society.]

5a–H. The dihedral angle between the two axial benzoate groups decreases relative to compounds in Class I.

Class III. No splitting is observd in the CD spectrum of the 1,3-ee dibenzoate (**5**), since the two C–O bonds are coplanar.

Class IV. The *A* value of 16 for 1,3-ea dibenzoates (**6**, **7**) is smaller than those of the 1,2-ee and 1,2-ea units ($A=\pm62$), since it is inversely proportional to the square of the interchromophoric distance.

Class V.As depicted in Figure 6.6 in this group (**8–11**), the 6-benzoate is directed towards the "right".

Class VI. In this group (**12–14**), the compounds contain 4a-substituents and the 6-benzoate is directed towards the "left".

Table 3.1 summarises the observed and calculated *A* values for tri- and tetra-benzoates. When entry 1 is compared with the rest of the table, it becomes clear that 1,2,3-eee tribenzoate exhibits a very weakly split CD signal because the 1,2

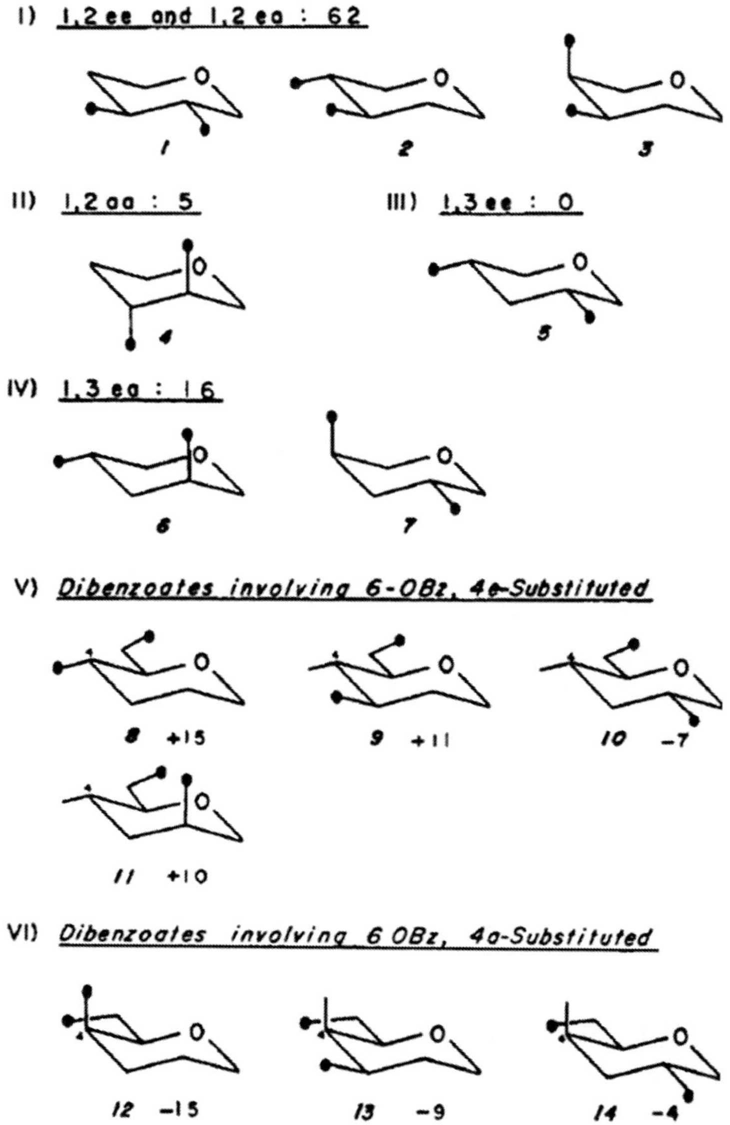

Figure 3.6 The *A* value of various pyranose *p*-bromobenzoates. The sign of the *A* values is dependent on the chiralities of the benzoate groups. Benzoate groups are indicated by hatched circles and e and a are used to denote equatorial and axial bonds. The anomeric carbon is normally α-OMe but occasionally β-OMe; the unsubstituted carbons at other locations are either methylenes or substituted with nonchromophoric groups such as –OAc, –OMe, $-CH_2OH$ and –Me. However, in the 6-benzoates (units **9–11**, **13**, **14**), the nonbenzoate substituents at C–4 are also depicted because the 6-benzoate conformation is affected by the C–4 substituent.[4] [Reprinted with permission from *J. Am. Chem. Soc.*, 1981, **103**, 5591. Copyright 1981, American Chemical Society.]

Table 3.1 Observed and calculated A ($\Delta\varepsilon_l - \Delta\varepsilon_r$) values of tri- and tetra-*p*-bromobenzoates.[4] [Reprinted with permission from *J. Am. Chem. Soc.* 1981, **103**, 5591. Copyright 1981, American Chemical Society.]

entry	*compound*[b]	236/253 nm	A (obsd)	A (calcd)[a]
1	β-Xyl-2e,3e,4e–	−1/+1	+2	0
2	β-Ara-2e,3e,4a–	−32/+101	+133	+140
3	α-Gal-2e,3e,4a–	−42/+95	+137	+140
4	α-Man-6-O-TBDMS[c]– 2a,3e,4e–	+29/−91	−120	−140
5	α-Glu-3-OMe-2e,4e,6–	−1/+8	+9	+8
6	α-Glu-2e,3e,6–	−22/+47	+69	+66
7	α-Glu-2-OMe-3e,4e,6–	+11/−22	−33	−36
8	α-Glu-2-OAc-3e,4e,6–	+10/−22	−32	−36
9	α-Gal-3 −OAc-2e,4a,6–	0/+6	+6	−3
10	α-Gal-2e,3e,6–	−14/+31	+45	+49
11	β-Gal-2e,3e,6–	−16/+30	+46	+49
12	β-Glu-2e,3e,4e,6–	−7/+25	+32	+19
13	α-Glu-2e,3e,4e,6–	−4/+20	+24	+19
14	β-Gal-2e,3e,4a,6–	−26/+74	+100	+112
15	α-Gal-2e,3e,4a,6–	−29/+70	+99	+112
16	α-Man-2a,3e,4e,6–	+23/−66	−89	−104

[a]Estimated A value from sum of component dibenzoate units 1–14, see Figure 3.6.
[b]Xyl, Ara, Gal, Man and Glu refer, respectively, to xylose, arabinose, galactose, mannose and glucose.
[c]*tert*-Butyldimethylsilyl.

and 2,3 amplitudes cancel each other out and the 1,3-ee amplitude is estimated to be nil (**5** in Figure 3.6). In the case of 1,2,3-eea tribenzoates (entries 2–4), the experimental values are large ($|A| = 120–140$) since the 1,2, 1,3 and 2,3 interactions have the same amplitudes. A reasonably accurate calculated value can be obtained for the sum of the three-component dibenzoate units based on the additivity relationship. For example, in the context of the second entry (β-Ara-2e,3e,4a–) + 62 (2e,3e; **1** in Figure 3.6) + 62 (3e,4a: **3** in Figure 3.6) +16 (2e,4a; **7** in Figure 3.6) = +140, while the observed value is +133.

Entries 5–11 of Table 3.1 are tribenzoates, which contain a 6-benzoate (6-OBz). The 6-OBz groups are orientated towards the "right" or "left" depending on whether the 4-substituent is equatorial or axial. Evidence for this is provided by the close agreement between the observed and calculated A values. For example, for entry 6 (2e,3e,6–): +62 (2e,3e; or unit **1**) − 7 (2e,6; or unit **10**) +11 (3e,6; or unit **9**) = +66, while the observed value is +69. This additivity still holds to a remarkable extent, even for the tetra-benzoates in entries 12–16, despite the fact that there are now six component interactions due to the 6-OBz group also being involved. For example, for entry 14 (2e,3e,4a,6–): + 62 (2e,3e; or unit **1**) + 16 (2e, 4a; or unit **7**) − 4 (2e,6; or unit **14**) +62 (2e,3e; or unit **3**) −9 (3e,6; or unit **13**) −15 (4a,6; or unit **12**) = +112, while the observed value is +100. The use of the additivity relationship to obtain the calculated CD intensities of some other compounds are provided in Figure 3.7.[5]

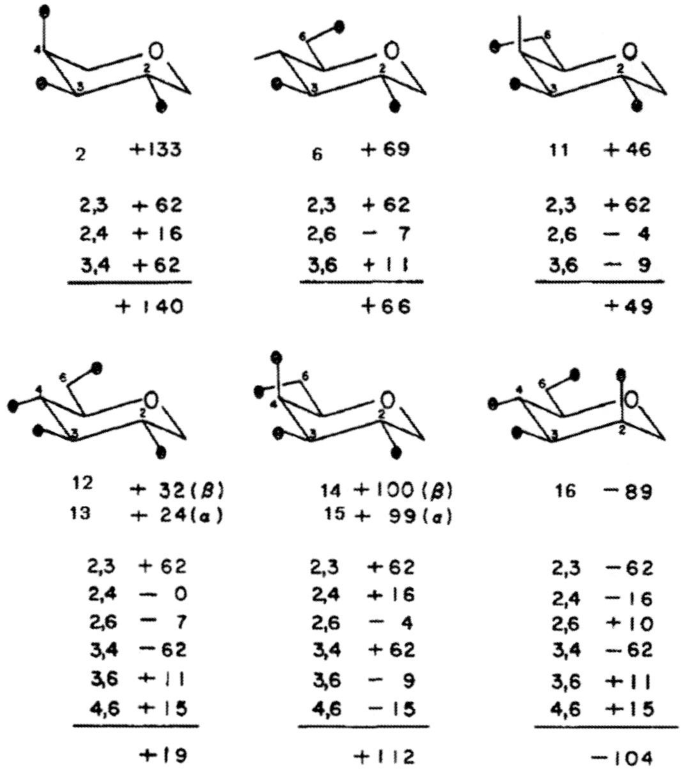

Figure 3.7 Observed and calculated *A* values of some *tris*- and *tetrakis*-(*p*-bromobenzoates). The numbers in bold face on the left-hand side immediately below each structure correspond to entry numbers in Table 3.1.[5] [Reprinted with permission from *J. Am. Chem. Soc.*, 1982, **104**, 1178. Copyright 1982, American Chemical Society.]

In summary, when applying the approach used in Figure 3.6, the values for the various units can be regarded as constants which are dependent only on the benzoate substitution pattern and are independent of the nature of the other substituents.

3.1.4 Determination of the Absolute Configuration of Oligonaphthalenes[6]

Porphyrins, particularly *meso*-tetraphenylporphyrins (H_2TPPs), have been used to extend the applicability of the exciton coupling method to configuration studies of molecules with remote stereogenic centres and to conformational studies of biopolymers.[7–9] Since porphyrins are rigid and have a very intense absorbance band ($\varepsilon = 150\,000$–$400\,000$) called the B (or Soret) band in the

400–420 nm region, at a much longer wavelength than those of most normal substrates, the analysis of large exciton-coupled CD usually provides unambiguous results. In the context of H_2TPPs, the edtms of the B transition are usually defined as being aligned along the N–H/N–H (B_x) and the N/N axes (B_y). However, because of the near degeneracy and approximate equivalence of these transitions, any linear combination of B_x and B_y could be used. The effective transition moments of mono-*p*-carboxylated H_2TPPs are believed to be aligned between opposite methine carbons, *i.e.* from the phenyl to mono-*p*-carboxyphenyl groups.[8] Two H_2TPP rings can be introduced onto the terminal naphthalene units of 1,1'-position-linked optically active naphthalene dimers and oligomers, Figure 3.8. The largest such system that has been studied was an octamer of 1,1'-binaphthyl, consisting of 16 naphthalene units.[6] According to the X-ray analysis of trimers and tetramers, the dihedral angles of the two naphthalene units ranged from 74.7 to 113°, with an average of 90.02°, while molecular mechanics calculations predicted dihedral angles of around 73.3–100.9° (average, 89.4°). Since the average dihedral angles appear to change by about 90° upon the introduction of each successive naphthalene unit, this is an ideal system for exploring the relationship between the chirality and CD intensity, Figure 3.9. For example, Figure 3.9A shows *(S,S,S)*-quaternaphthalene viewed from the top, where each side chain is helically arranged every 90° in a clockwise direction. As a result, the two H_2TPPs are orientated in an anticlockwise direction, which leads to a negative exciton-coupled CD spectrum, Figure 3.9B. However, in *(S,R,S)*-quaternaphthalene, the two H_2TPPs are aligned in a clockwise direction, leading to a positive split CD signal, Figure 3.9C.

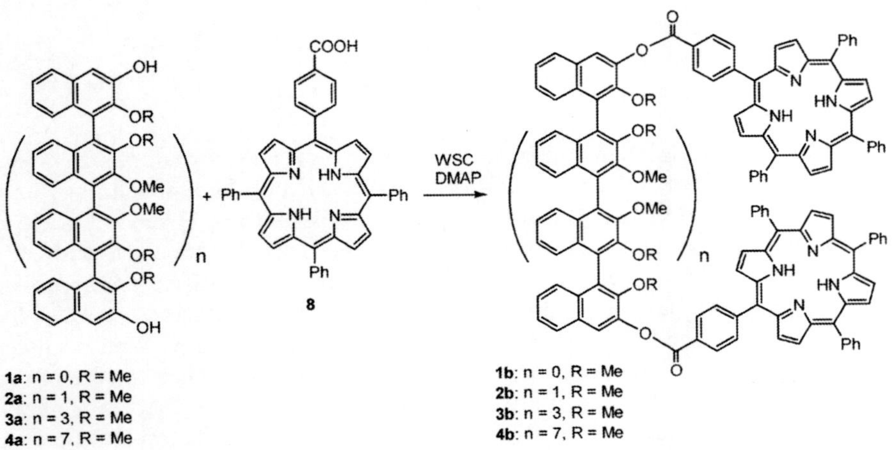

Figure 3.8 Synthesis and structures of oligonaphthalene linked to two H_2TPP.[6] [Reprinted with permission from *Org. Lett.*, 2006, **8**, 2587. Copyright 2006, American Chemical Society.]

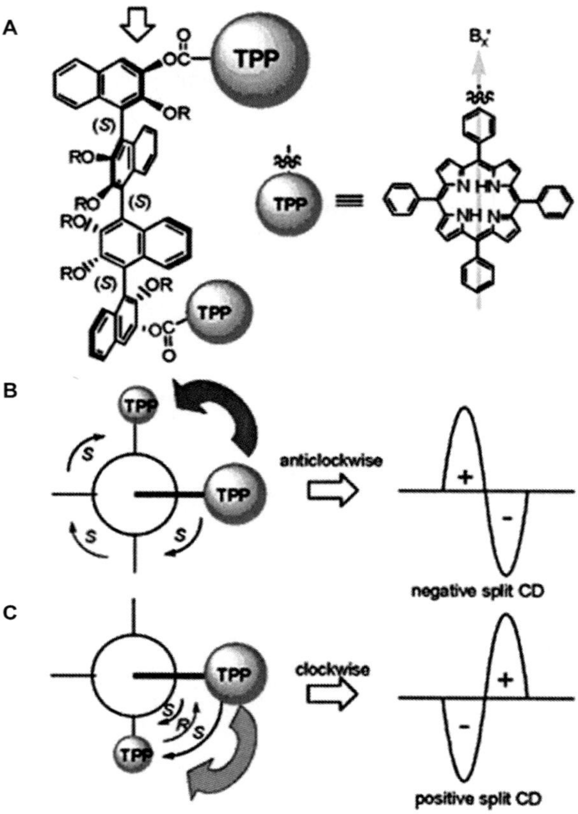

Figure 3.9 Application of the exciton chirality method to the oligonaphthalene skeleton.[6] [Reprinted with permission from *Org. Lett.*, 2006, **8**, 2587. Copyright 2006, American Chemical Society.]

Figure 3.10 contains the absorption and CD bands of compounds **1b–4b** in Figure 3.8. The B bands (*ca.* 420 nm) have almost the same morphology and intensity in the electronic absorption spectra, while in the CD spectra, a clear splitting of the CD curves is observed and the sign patterns are consistent with the assumption described in Figure 3.9. For example, negative and positive exciton-coupled CD signals are anticipated for *(S,S,S)*-**2b** and *(S,R,S)*-**2b**, respectively, Figures 3.9B and C, as is observed experimentally. Well resolved CD bands are detected, even for the diastereomers of the 16mer **4b**, where the estimated distance between the oxygen atoms on the top and bottom naphthalenes is 66 Å. Because the ratio of the distance between the top and bottom H_2TPP moieties of oligonaphthalenes **2b**(4mer)/**3b**(8mer)/**4b**(16mer) is about 0.5 : 1 : 2 and the total amplitudes of the CD signals are inversely proportional to the square of the interchromophoric distance, the estimated ratio of the CD

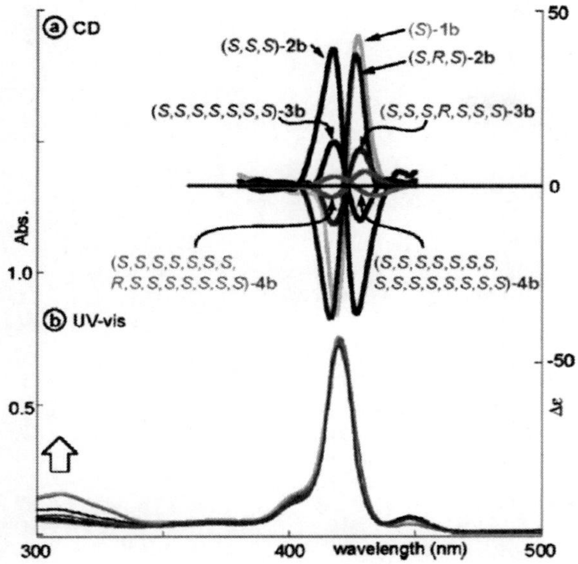

Figure 3.10 CD (a) and absorption (b) spectra of **1b–4b**.[6] [Reprinted with permission from *Org. Lett.*, 2006, **8**, 2587. Copyright 2006, American Chemical Society.]

amplitudes is 4 : 1 : 0.25, respectively, very close to the observed values of 75 : 22 : 6 (=3.4 : 1 : 0.27).

3.1.5 Determination of the Absolute Configuration of a Natural Catechol Product, Haematoxylin, using the Phthalocyanine Chromophore[9]

Haematoxylin, a natural catechol, exhibits an absorption band at wavelengths shorter than 300 nm, since the catechol molecules are the only chromophores that are present in the sample. Analysis of the CD spectra is somewhat problematic, since the absorption band lies close to the normal measurement limits. Two phthalocyanine (Pc) chromophores with strong absorption near 700 nm were, therefore, introduced at the two catechol moieties, Figure 3.11, and the absorption and CD spectra of the resulting compound were analysed. Although two kinds of chromophore are included in the product molecule, the interaction between the catechol and phthalocyanine can essentially be neglected, since their transition energies are quite different (*ca.* 250 and 700 nm, respectively). According to geometry-optimisation calculations, two structures designated as structures A and B in Figure 3.12 can be considered for a *(*6a*S*,11b*R)* absolute configuration, depending on the orientation of the CH_2 carbon in the central pyran ring,

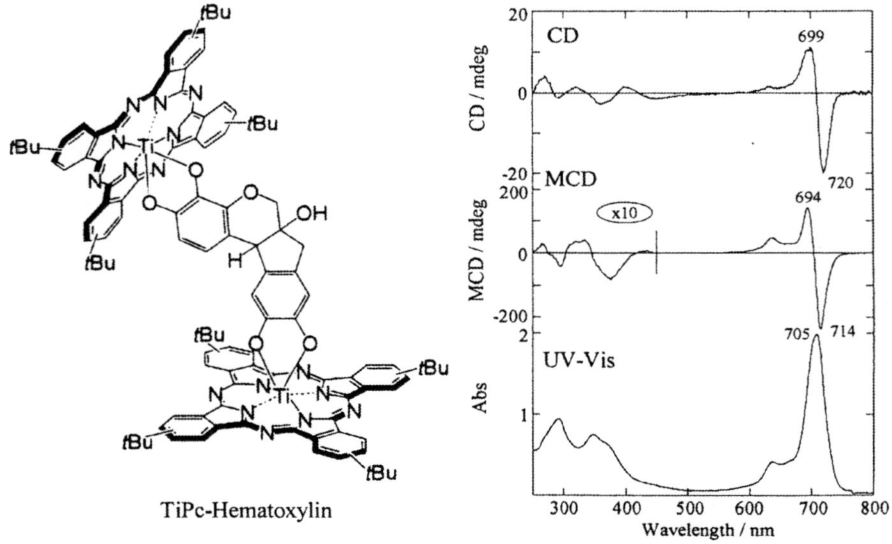

Figure 3.11 Haematoxylin and TiPc-linked haematoxylin (left) and electric absorption, magnetic circular dichroism (MCD) and CD spectra (right) of the latter.[9]

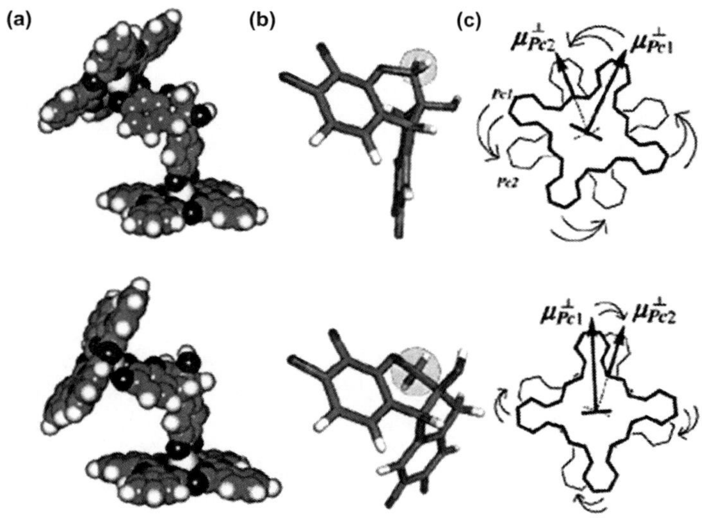

Figure 3.12 Two possible low-energy conformers of TiPc-(6a*S*,11b*R*)-haematoxylin obtained from DFT calculations (structure A, top; Structure B, bottom). (a) Optimised geometries. (b) Haematoxylin sites in the two optimised structures. The circle indicates the flexible CH_2 group in the pyran ring. (c) Schematic representations of supramolecular chirality.[9]

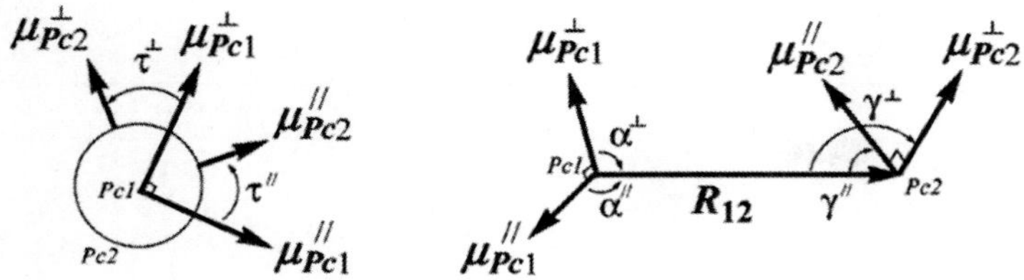

Figure 3.13 Geometrical parameters for CD analysis of systems containing two phthalocyanine (Pc) systems: $\boldsymbol{\mu}$ indicates the electronic dipole transition moment; $\perp$and // indicate whether the $\boldsymbol{\mu}$ is normal or parallel, respectively, to the catechol plane linked to a Ti atom.[9]

Figure 3.12(b). If the two Pcs are labelled as Pc1 (near side) and Pc2 (far side), the configuration is left-handed in structure A and right-handed in structure B, respectively, when viewed from Pc1 to Pc2. Since the calculated energy difference between structures A and B is negligible, the CD analysis was carried out under the assumption that these two structures exist in a 1 : 1 ratio. The alignments of the edtms on the Pc rings are depicted in Figure 3.13. The normal and parallel suffixes indicate whether the edtm is normal or parallel to the catechol plane connected to the Ti atom. The presence of multiple transitions polarised along different directions is confirmed by the presence of CD bands of opposite sign in the Q band region, Figure 3.11.

The analysis was carried out using the exciton coupling theory for degenerate coupled-oscillator systems.[10] The point-dipole approximation was applied, equations (1.9) and (1.10), and the interaction energy between the two Pc chromophores was predicted to be inversely proportional to the cube of the interchromophoric distance (R_{12}) and proportional to the square of the transition moment ($\boldsymbol{\mu}_e$) of the interacting chromophore on this basis. This approach is only valid when the separation of the chromophores is much larger than the transition dipole length. In this system, the application of the approximation is clearly justified, since the distance between the centres of gravity of the TiPc chromophores (R_{12}) is *ca.* 13 Å, while in contrast the transition dipole length calculated from the band fits is between 0.8 and 0.9 Å. When the two degenerate coupled-oscillator systems of the TiPc–haematoxylin complex are taken into consideration, the exciton band energies, $\varepsilon^{\perp}(\pm)$ and $\varepsilon^{//}(\pm)$, and the rotatory strengths, $R^{\perp}(\pm)$ and $R^{//}(\pm)$, can be written based on the following four equations using several angles defined in Figure 3.13 and values listed in Table 3.2, which were obtained using DFT optimised structures:

$$R^{\perp}(\pm) = \pm[\{\varepsilon_{Pc}{}^{\perp}(\boldsymbol{\mu}_{Pc}{}^{\perp})^2 R_{12}\}/(4\hbar)]\sin\alpha^{\perp}\sin\gamma^{\perp}\sin\tau^{\perp} \tag{3.1}$$

$$\varepsilon^{\perp}(\pm) = \varepsilon_{Pc}{}^{\perp} \pm [(\boldsymbol{\mu}_{Pc}{}^{\perp})^2/4\pi\epsilon_0(R_{12})^3](\sin\alpha^{\perp}\sin\gamma^{\perp}\cos\tau^{\perp} + 2\cos\alpha^{\perp}\cos\gamma^{\perp}) \tag{3.2}$$

Table 3.2 Geometrical parameters for the TiPc–(6a*S*, 11b*R*)–haematoxylin complex.[9]

structure		α	β	γ	R_{12} / Å
A	$\mu Pc^{//}$	81.96	84.20	40.03	13.20
	$\mu Pc^{\perp}$	61.00	65.29	38.23	13.20
B	$\mu Pc^{//}$	62.75	90.01	351.70	13.01
	$\mu Pc^{\perp}$	66.97	61.17	339.00	13.01

Table 3.3 Results of coupled oscillator analysis for TiPc–(6a*S*, 11b*R*)–haematoxylin and Gaussian fit analysis of the observed CD spectrum for TiPc–haematoxylin.[9]

		energy (cm^{-1})	*wavelength* (nm)	*oscillator strength*	*rotatory strength*/ ($D\mu_B$)
structure A	$\varepsilon^{//}(-)$	14 078	710.3	0.0337	−3.844
	$\varepsilon^{//}(+)$	14 149	706.7	0.2358	3.844
	$\varepsilon^{\perp}(-)$	14 299	699.3	0.0653	−2.630
	$\varepsilon^{\perp}(+)$	14 378	695.5	0.1600	2.630
structure B	$\varepsilon^{//}(-)$	14 075	710.5	0.0156	0.783
	$\varepsilon^{//}(+)$	14 153	706.5	0.2439	−0.783
	$\varepsilon^{\perp}(-)$	14 295	699.5	0.0491	1.547
	$\varepsilon^{\perp}(+)$	14 382	695.3	0.1762	−1.547
Gaussian fit		13 954	716.7	0.0927	−0.935
		14 245	702.0	0.7750	0.746

$$R^{//}(\pm) = \pm[\{\varepsilon_{Pc}{}^{//}(\boldsymbol{\mu}_{Pc}{}^{//})^2 R_{12}\}/(4\hbar)]\sin\alpha^{//}\sin\gamma^{//}\sin\tau^{//} \tag{3.3}$$

$$\varepsilon^{//}(\pm) = \varepsilon_{Pc}{}^{//} \pm [(\boldsymbol{\mu}_{Pc}{}^{//})^2/4\pi\epsilon_0(R_{12})^3](\sin\alpha^{//}\sin\gamma^{//}\cos\tau^{//} + 2\cos\alpha^{//}\cos\gamma^{//}) \tag{3.4}$$

It should be noted that the angle τ is always taken in an anticlockwise direction when the system is viewed from Pc1 to Pc2 and that the notations plus and minus denote in-phase and out-of-phase transitions, respectively. The results of the coupled oscillator analysis are summarised in Table 3.3 and shown in Figure 3.14. In this calculation, the absolute configuration was arbitrarily chosen to be the (6a*S*, 11b*R*) form.

Because of the very slight splitting (30–40 cm^{-1}) of these exciton states, the observed CD intensities are considerably weaker than the actual intensity of the component bands of the couplet. It is obvious that the signs of the CD signals for the two lowest-energy conformers are opposite and the CD intensity of the "structure A" conformer is approximately 3 times larger than that of the "structure B" conformer. Calculated absorption and CD spectra were generated by superimposing the spectra of the two conformers for the (6a*S*, 11b*R*)-form in a 1 : 1 mixture, Figure 3.14, right. Since the calculated spectrum of the

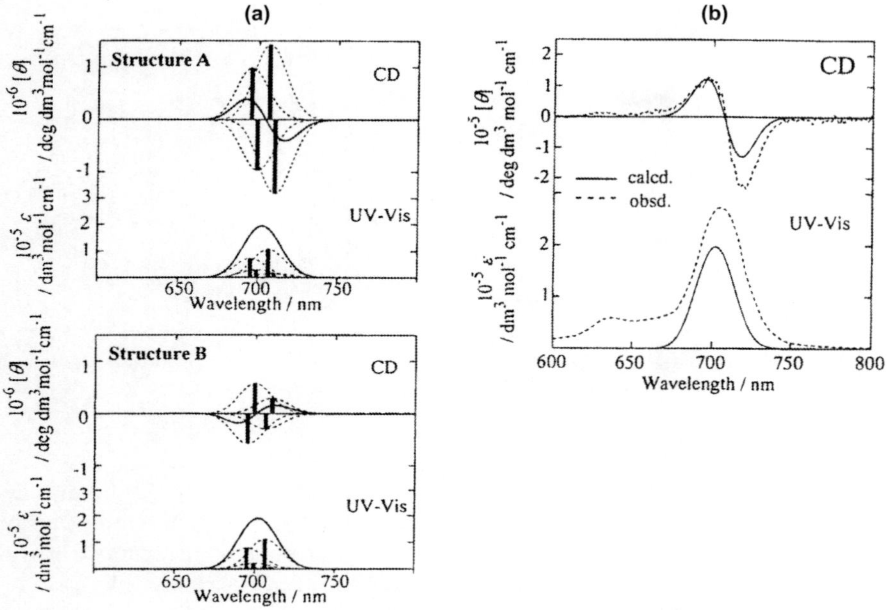

Figure 3.14 Calculated electronic absorption and CD spectra of the two conformers in the TiPc–(6a*S*, 11b*R*)–haematoxylin complex (left). Each band was described as a single Gaussian curve (broken line). The solid line indicates the superposition of all Gaussian curves. Comparison between the experimental spectra of *tert*-butylated TiPc–hematoxylin and the calculated spectra (solid line) generated by superimposing the spectra of the two conformers for the (6a*S*, 11b*R*)-form in a 50 : 50 mixture (right).[9]

(6a*S*, 11b*R*)-form is in close agreement with the observed CD signals, while in contrast that calculated for the (6a*S*, 11b*R*)-form is the mirror image of the observed CD spectrum, the absolute stereochemistry for haematoxylin was assigned as being (6a*S*, 11b*R*)

3.1.6 Supramolecular Chirality in a Bisporphyrin System Axially Coordinated by a Chiral Guest[11]

The ethane-bridged *bis*-octaethylporphyrin zinc complex in Figure 3.15 is an ideal achiral host for carrying out comprehensive investigations of various aspects of supramolecular chirogenesis based on its interaction with different chiral guest compounds. Of the compounds studied, bidentate ligands are particularly important because of their ability to form rigid, stable tweezer-type complexes with extremely strong optical activity. However, key questions relating to the exact molecular structure and the true origin of the induced chirality initially remained unsolved. In order to resolve this problem,

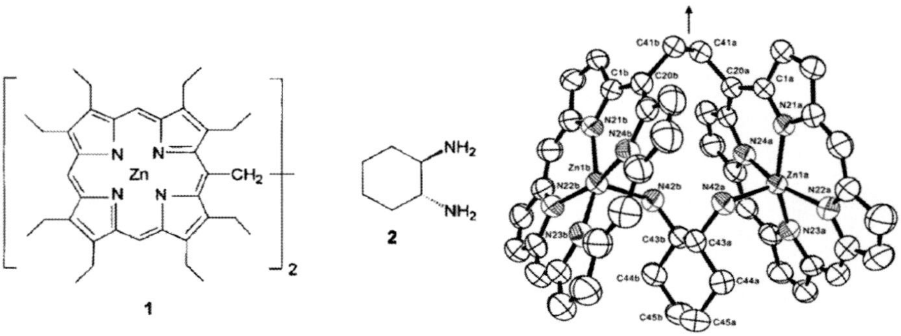

Figure 3.15 A *bis*octaethylporphyrin linked by a *meso-meso* ethano bridge (left) and its X-ray structure linked by (*R,R*)-1,2-diaminocyclohexane (right).[11]

the structure of a *bis*-octaethylporphyrinatozinc(II)-(*R,R*)-1,2-diaminocyclohexane tweezer-type complex was analyzed by X-ray crystallography. Subsequently, the structure was examined in solution by electronic absorption, magnetic circular dichroism (MCD) and CD spectroscopy, Figure 3.16, and theoretical calculations were carried out using the coordinates obtained in the solid state. There are three well-resolved absorption bands at 410, 419 and 436 nm in the B (or Soret) band region of the electronic absorption spectrum of the porphyrin. The centres of the MCD bands coincide very closely with those observed in the absorption spectrum due to the orbitally nondegenerate nature of the excited states (Gaussian-shaped Faraday $\mathcal{B}_0$ terms dominate the MCD spectra of low symmetry compounds). Close-lying excited states associated with bands with differing polarisations tend to be mixed in an applied field in a manner which results in coupled pairs of oppositely signed Faraday $\mathcal{B}_0$ terms, see Figure 1.23. The two positively signed CD bands at 410 and 419 nm can be assumed to have different polarisations, therefore, since positive and negative $\mathcal{B}_0$ terms are observed in the MCD spectrum at 408 and 424 nm, respectively.

This provides an ideal example for exciton coupling analysis, since the (*R,R*)-1,2-diaminocyclohexane coordinating to the Zn(II) ion is not a chromophore so the stealing of CD intensity from the ligand transitions can be safely neglected. As shown in Section 3.1.5 above, this tweezer system consists of two identical porphyrin moieties, each having two different transition moments $B^{//}$ and $B^{\perp}$, which are aligned in parallel and perpendicular orientations relative to the ethane bridge. This rearrangement results in two major types of degenerate exciton interaction ($B^{//} - B^{//}$ and $B^{\perp} - B^{\perp}$ couplings), leading to four exciton states. The spectral parameters for the $B^{//}$ and $B^{\perp}$ transitions of the porphyrin monomer were first obtained from a spectral band deconvolution analysis for zinc octaethylporphyrin with a $-CH_2OC_2H_5$ substituent at the *meso*-position ($B^{//}$: exciton energy = 23 743 cm^{-1}, bandwidth = 576 cm^{-1}; magnitude of the transition moment = 7.428 D; $B^{\perp}$:

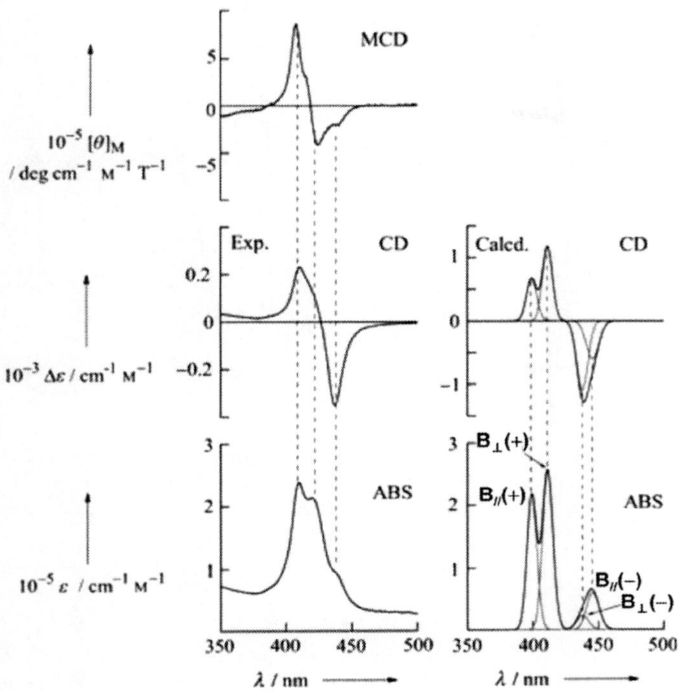

Figure 3.16 Experimental MCD, CD and absorption spectra recorded in CH_2Cl_2 (left), as well as the calculated CD and absorption spectra (right) of the B band region of the tweezer. Each calculated band is described by a single Gaussian curve. The notations plus and minus represent the in-phase and out-of-phase transitions, respectively.[11]

exciton energy = 23 575 cm^{-1}; bandwidth = 525 cm^{-1}; magnitude of the transition moment = 7.118 D). The values for the exciton interaction energies were then derived from these parameters ($B^{//}$–$B^{//}$: 1300 cm^{-1}, $B^{\perp}$–$B^{\perp}$: 740 cm^{-1}, $B^{//}$–$B^{\perp}$: 250 and 225 cm^{-1}). The contribution arising from non-degenerate couplings ($B^{//}$–$B^{\perp}$) was calculated to have only 1% of the intensity of the degenerate couplings ($B^{//}$–$B^{//}$ and $B^{\perp}$–$B^{\perp}$) so its contribution was neglected in the calculation. The polarisations of the two major calculated bands (399 and 411 nm) are perpendicular to each other. As observed in Figure 3.17, the relative orientation between these two pairs of electronic transitions is anticlockwise. The exciton chirality method predicts that negative chirality should be observed experimentally, Figure 3.16, under these circumstances. The calculated absorption spectrum has three well-resolved bands that are broadly similar to the experimental results shown in Figure 3.16. The face-to-face conformation and short distance between the centres of the interacting electric dipoles (r_{12} = 6.23 Å) results in the higher energy ($B^{//}(+)$ and $B^{\perp}(+)$) and lower energy ($B^{|}(-)$ and $B^{\perp}(-)$) exciton states being

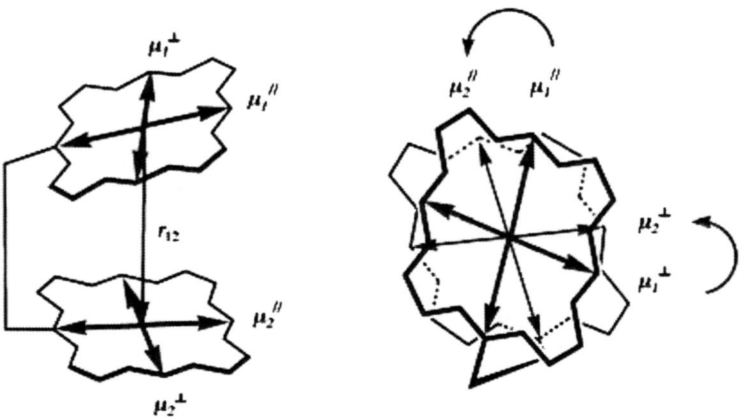

Figure 3.17 Definitions of the direction of the transition dipole moments used in the theoretical analysis: r_{12} is the distance between the centres of gravity of the porphyrin units; the subscripts 1 and 2 (for $\boldsymbol{\mu}$) are used to distinguish between the two chromophores.[11]

well separated. These pairs of states become optically allowed and partially forbidden, respectively, due to the strong exciton interactions. The MCD $B^{//}(+)$ and $B^{\perp}(+)$ bands should theoretically be of opposite sign because of their orthogonal polarisations. The experimental results are fully consistent with this. A minus-to-plus CD pattern is anticipated for the $B^{//} - B^{//}$ and $B^{\perp} - B^{\perp}$ couplings and this results in one intense negative CD signal at lower energy and two partially resolved positive CD signals at higher energy, in close agreement with what is observed experimentally, Figure 3.16. The main features in the CD spectrum of the *bis*-octaethylporphyrinatozinc(II)-(*R,R*)-1,2-diamino-cyclohexane tweezer-type complex can, therefore, be fully explained by exciton coupling theory.

3.1.7 A Chiral Biscyanine Dye exhibiting an Exciton Couplet with Well-resolved Absorption Bands[12]

A neutral biscyanine dye, Figure 3.18, was synthesised based on a condensation reaction of (1*S*,2*S*)-(+)-*trans*-1,2-cyclohexane-1,2-diamine and 7-piperidinohepta-2,4,6-trienal (merocyanine). Stepwise protonation with trifluoroacetic acid results in the formation of mono- and di-cationic species, Figure 3.18A and B. A set of isosbestic points is observed. As shown in Figure 3.19, the neutral form has a broad absorption at 383 nm and typical exciton-split bisignate CD curves, which appear to be approximately consistent with the "sum rule". Since the sign of the lower energy CD band is negative, the exciton chirality is negative (*i.e.* the two transition moments have an anticlockwise screw sense), see Figure 1.16. The monocationic form

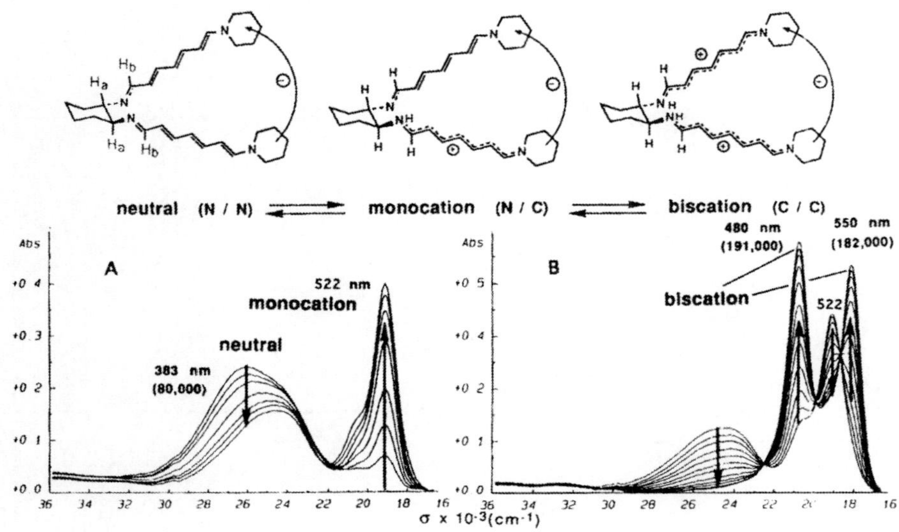

Figure 3.18 Structures of a biscyanine dye and its monocationic and dicationic forms obtained by adding trifluoroacetic acid (TFA) (top). Absorption spectroscopic changes of the neutral dye to its (A) monocation and subsequently its (B) *bis*cation observed by adding TFA.[12b] [Reprinted with permission from *J. Am. Chem. Soc.*, 1993, **115**, 4769. Copyright 1991, American Chemical Society.]

displays a narrow band at 522 nm and a broad band at 405 nm, arising from cationic and neutral cyanines, respectively (this is, therefore, a non-degenerate coupled-oscillator system). Negative and positive envelopes of CD intensity were observed, corresponding to these absorption bands. Although the intensities of these CD bands appear to be quite different based on the relative $\Delta\varepsilon_{max}$ values, the actual integrated intensities are comparatively similar as described in the inset, which provides the rotatory strength values, R_1 and R_2. The spectra of the dicationic species are characterised by two intense absorption peaks at 550 and 480 nm and negatively and positively signed CD bands. The possible reasons for the presence of these two absorption peaks were carefully considered. The existence of two species absorbing at different wavelengths was ruled out because the titration curves plotted at 550 and 480 nm were identical, Figure 3.19, indicating the presence of a single species. An exciton coupling of a *bis*cation species was supported by the following features: 1) The two peaks at 550 and 480 nm are equidistant from that of the monocyanine (515 nm), and 2) the CD spectrum contains bisignate CD bands of similar intensity. The reasons for the appearance of well-resolved absorption peaks were explored, since the exciton couplets of small molecules are usually not as well resolved as this. Two explanations were provided. Firstly, although the energy gap between the two bands is relatively large (2890 cm^{-1}), the bands are located in the long wavelength region, where the

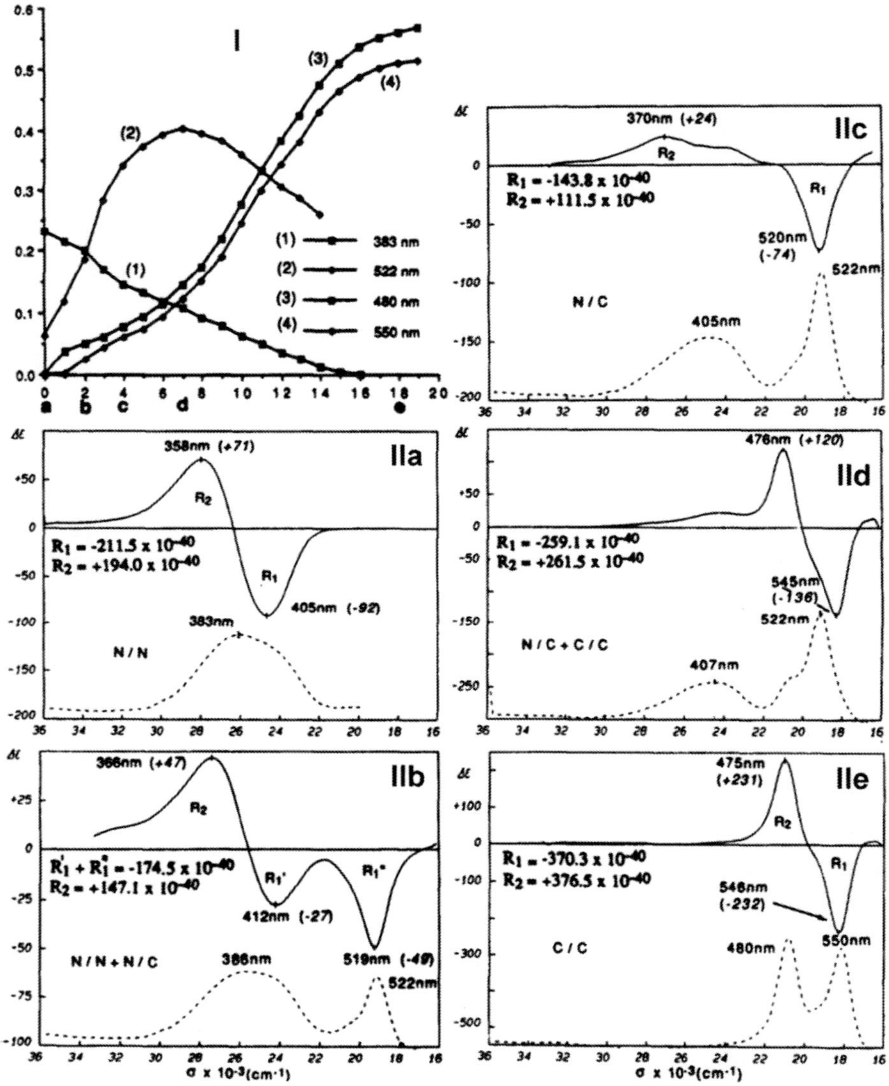

Figure 3.19 (I) Titration curves of neutral Schiff base with TFA in CH_2Cl_2. Horizontal axis: Each aliquot represents 1.56×10^{-3} mmol of TFA (5×10^{-4} M) solution added to 1.3 mL of neutral Schiff base: (a–e) relevant stages of the titration with TFA. Vertical axis: absorbance at λ_{max}: (1) neutral, 383 nm, (2) monocation, 522 nm, (3) *bis*cation, 480 nm and (4) *bis*cation, 550 nm. (II) Changes in CD (solid line) and absorption (dashed line) spectra of neutral Schiff base accompanying the titration with TFA, both in CH_2Cl_2. IIa–IIe correspond to stages a–e in (I): (a) neutral Schiff base; (b) mixture of neutral Schiff base and monocation species; (c) monocation; (d) mixture of monocation and *bis*cation; (e) *bis*cation species.[12b] [Reprinted with permission from *J. Am. Chem. Soc.*, 1993, **115**, 4769. Copyright 1991, American Chemical Society.]

energy splitting is emphasised by the wavelength scale. Secondly, the absorption bands have very narrow half-bandwidths of *ca.* 1520 cm^{-1} which are much smaller than those normally observed in the UV region in the spectra of small molecules ($\sim$5000 cm^{-1}).

3.1.8 CD and Absolute Configuration of C_3 Symmetry Chiral Cyclotriveratrylenes[13,14]

Cyclotriveratrylenes are highly suitable for exciton coupling analyses, because three identical benzene chromophores are rigidly arranged in a "crown" conformation. The centre-to-centre distance of the two adjacent benzene rings is *ca.* 4.79 Å and that of the two adjacent CH_2 groups is *ca.* 2.0 Å, while the benzene planes are tilted from the C_3 axis by 43°. When two substituents R' and R are linked to each benzene ring, Figure 3.20, the derivative becomes optically active. In analysing the spectra of cyclotriveratrylenes, the polarisation of the constituent benzene units has to be determined. When the R' and R substituents are –H, –OH, –O–alkyl, –O–acyl and/or –Br groups, they exhibit a relatively weak short axis polarised ^1L_b band at *ca.* 290 nm (B_{2u} excited state) and a relatively intense long axis polarised ^1L_a band in the 230–240 nm region (B_{1u} excited state), Figure 3.21. Since R' and R are different, the presence of two substituents on each benzene ring causes a slight rotation of the ^1L_b (θ_2) and ^1L_a (θ_1) edtms away from the short and long axes. The sense and the extent of the rotation depends primarily on the relative magnitude of the spectroscopic moments of R' and R, Figure 3.20. It has been assumed[15] that θ_2 (associated with the band at longer wavelength) can be evaluated based on vector addition of the relevant substituent spectroscopic moments (a substituent group parameter which determines band intensities in the spectra of aromatic compounds),[16,17] and that θ_1 is always roughly perpendicular to θ_2. In the case of mono-substituted compounds, this is

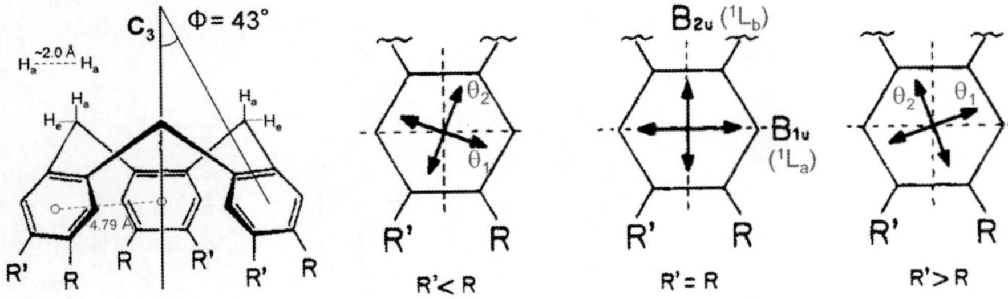

Figure 3.20 Summary of geometric and structural properties of (left) and the exciton mechanism in (right), C_3-cyclotriveratrylenes. Polarisation of the ^1L_b (θ_2) (long-axis polarised) and ^1L_a (θ_1) (short-axis polarised) transitions as a function of the spectroscopic moments of substituents R and R'.[13] [Reprinted with permission from *J. Am. Chem. Soc.*, 1981, **103**, 204. Copyright 1981, American Chemical Society.]

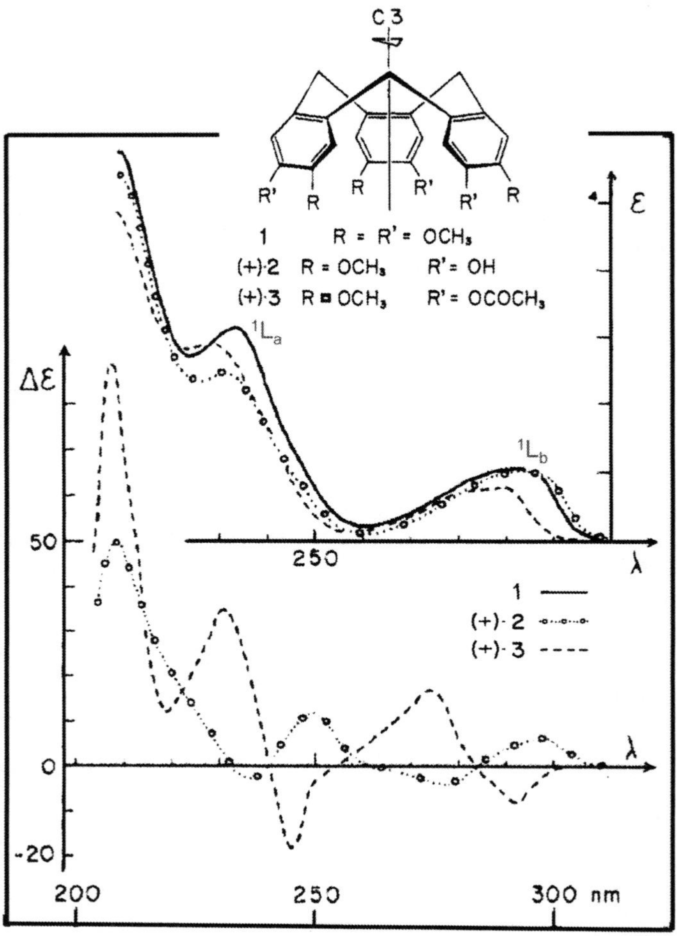

Figure 3.21 Isotropic absorption and CD spectra of **1**, (+)-**2** and (+)-**3** in methanol.[13] [Reprinted with permission from *J. Am. Chem. Soc.*, 1981, **103**, 204. Copyright 1981, American Chemical Society.]

proportional to the square of the transition moment induced by the substituent and additional intensity is observed in the ^1L_b band. This was researched in detail more than 30 years ago.[18] The polarisation direction of the ^1L_b transition, which is slightly rotated from the short B_{2u} axis, can be readily derived from molecular orbital calculations using commercially available software.

When R' = CH_3COO^- and R = CH_3O^-, θ is small and positive in value. The in-phase A-coupling state lies at higher energy than the out-of-phase E-coupling state and results in a band with a positive rotational strength. This

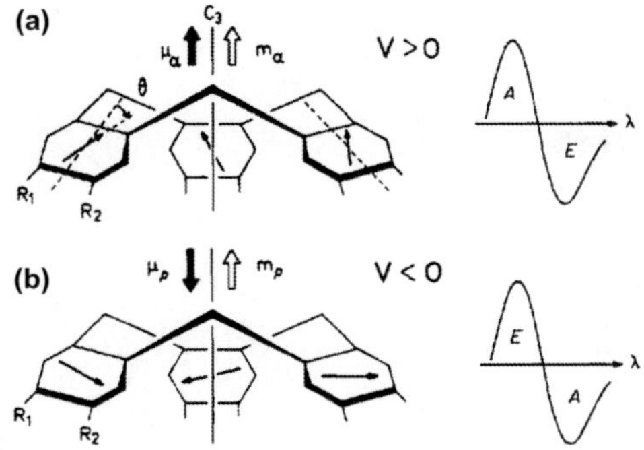

Figure 3.22 In-phase A-coupling for a small positive value of θ of the B_{2u} edtm in (+)-**3** (top left) results in positive rotational strength at higher energy (top right). The out-of-phase E-coupling can be obtained by inverting one or two arrows. Negative rotational strength is observed at lower energy (bottom right) for the in-phase A-coupling of the B_{1u} edtm (bottom left). [Reprinted with permission from *J. Am. Chem. Soc.*, 1985, **107**, 1299. Copyright 1985, American Chemical Society.]

can be explained using Figure 3.22. It can easily be rationalised on the basis of the differing coulombic interactions of the edtms that the A-coupling is higher in energy than the E-coupling, because in the former case, the tips or origins of the arrows of the edtms lie in close proximity and are of the same sign, while in the latter case, the tips and origins of adjacent arrows of electric transition moments lie in close proximity but are of opposite sign. The in-phase A-coupling of the three transition dipoles generates parallel combined electric ($\boldsymbol{\mu}$) and magnetic ($\boldsymbol{m}$) transition moments along the three-fold axis, giving rise to a positive rotational strength based on the scalar product of $\boldsymbol{\mu}$ and $\boldsymbol{m}$ in the Rosenfeld equation, equation (1.17). When the spectroscopic moment of R' is larger than that of R, such as when R' = OH and R = CH_3O^-, the CD signal becomes slightly negative. In this case, the direction of $\boldsymbol{m}$ is opposite to that of $\boldsymbol{\mu}$, so $\boldsymbol{\mu} \cdot \boldsymbol{m}$ becomes negative and the CD band at higher energy is negatively signed, Figure 3.21.

The CD signal of the long axis-polarised ^1L_a band between 230 and 240 nm (B_{1u} excited state) can be interpreted in a similar manner, Figure 3.22. When R' = CH_3COO^- and R = CH_3O^-, the value of θ is slightly positive. In-phase A-coupling for the B_{1u} symmetry transition generates antiparallel combined electric and parallel magnetic transition moments along the C_3 axis, so the observed rotational strength is negatively signed. The coulombic interaction between the signs associated with the tips or origins of the edtms is attractive in this context so the band associated with A-coupling lies at lower energy. As a result, it has been demonstrated based on a study of compounds with several different sets of R' and R groups that the CD signals that would be anticipated on this basis, Figure 3.23, are observed in the ^1L_b and ^1L_a band regions.[19]

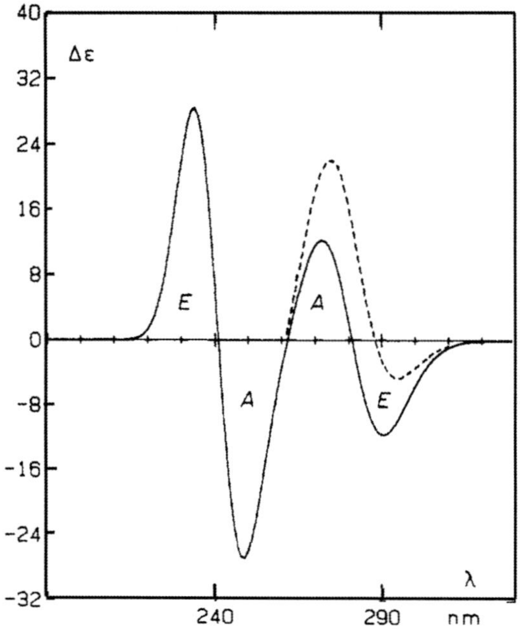

Figure 3.23 Theoretical sequence of exciton bands for small θ ($0 < \theta < 45°$): zero-order spectrum (solid line); after configuration interaction (broken line).[19] The CD curves were calculated using dipole strength values of 3×10^{-36} and 2×10^{-36} cgs, respectively, for the transitions to the B_{1u} and B_{2u} excited states with $\theta = +25°$. This corresponds to (*P*)-(+)-**3** in Figure 3.21. [Reprinted with permission from *J. Am. Chem. Soc.*, 1985, **107**, 1299. Copyright 1985, American Chemical Society.]

3.2 Polymer Systems (Systems Consisting of more than Four Chromophores)

Although a wide range of synthetic homopolymer systems have been reported very few are suitable for study by CD spectroscopy.

3.2.1 Helical Structures of N-Alkylated Poly(*p*-benzamide)s[20]

The synthetic polyamide depicted in Figure 3.24 appears to be the best available system for describing the application of exciton coupling theory to synthetic homopolymers. The constituting monomer unit, N-alkylated *p*-aminobenzoic acid, has a strong long-axis-polarised absorption band at *ca.* 300 nm, well separated from the weak short-axis-polarised band at *ca.* 230 nm, Figure 3.25. The spectrum of the polymer contains an absorption band (in $CHCl_3$) or shoulder (in CH_3CN) at *ca.* 300 nm and a band at *ca.* 260 nm,

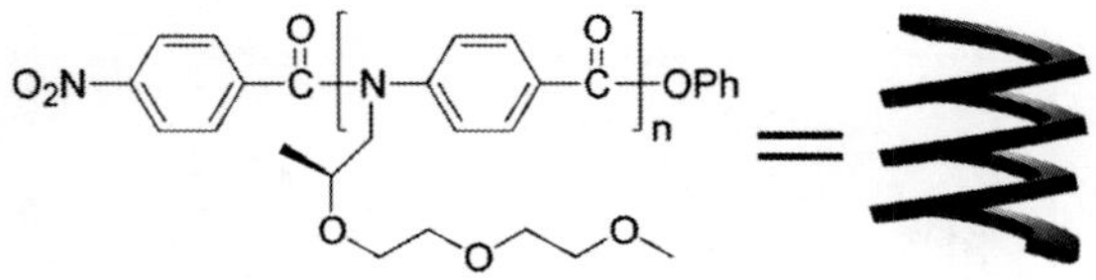

Figure 3.24 Structure of polyamide **1**.[20]

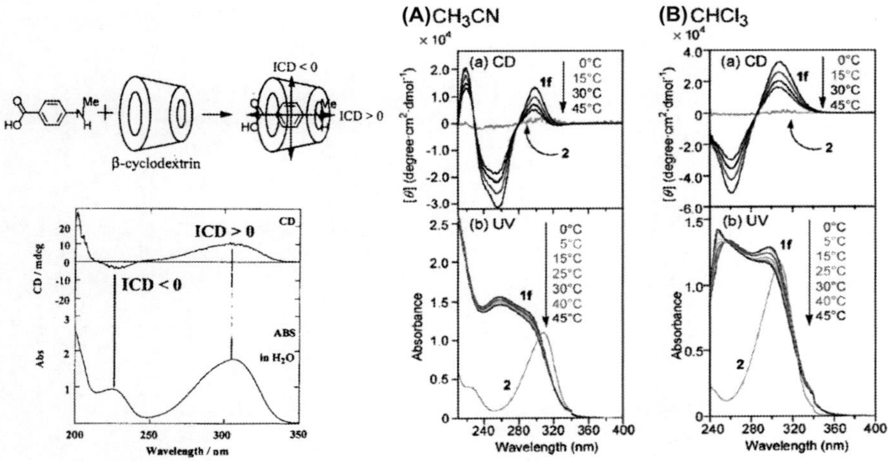

Figure 3.25 CD and electronic absorption spectra (left) of 4-(methylamino)benzoic acid in water in the presence of β-cyclodextrin. CD and absorption spectra (centre and right) of polymer **1** (polymerisation degree = *ca.* 25) in (A) CH_3CN and (B) $CHCl_3$ at various temperatures and those of the constituting monomer **2** (the monomer spectra were recorded at 15 °C). Assembled from 20 and 26.

with positively and negatively signed CD envelopes, respectively, Figure 3.25. The CD intensity increases with increasing molecular weight (MW) up to *ca.* 7000 amu (polymerisation degree = *ca.* 25), then saturates. The signal decreases with increasing measurement temperature. Oligomers of n = 2–8 were prepared to provide model complexes, which can be used to derive the key information required to accurately assign the spectra of the polymers. The spectrum of the n = 2 dimer contains absorption bands at *ca.* 320 and 280 nm, while those of the n = 3–8 oligomers contain a shoulder at *ca.* 300 nm and a peak in the 260–280 nm region, Figure 3.26. These spectra can be readily rationalised on the basis of exciton coupling theory, since the X-ray analyses of these structures indicate that three units combine to form one turn of a helix. If, as is likely, this is also the case in solution, the 320 and 280 nm peaks of the *cis*-dimer can be assigned to transitions from the groundstate to the lower and higher energy exciton states produced by in-phase and out-of-phase

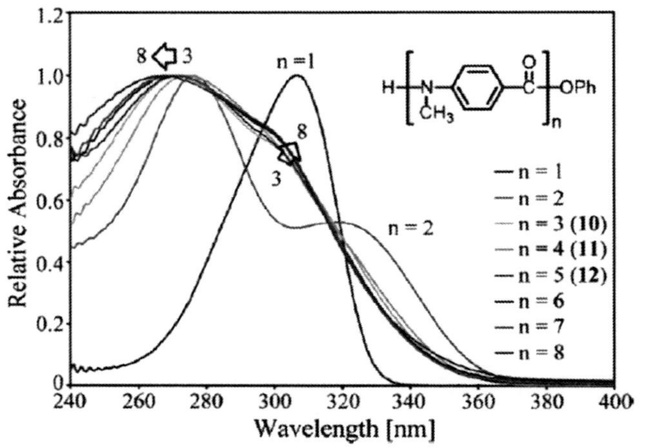

Figure 3.26 Room temperature electronic absorption spectra of 4-(methylamino)-benzoic acid oligomer phenyl esters in $CHCl_3$. The absorption intensity is normalised at the strongest peak.[20]

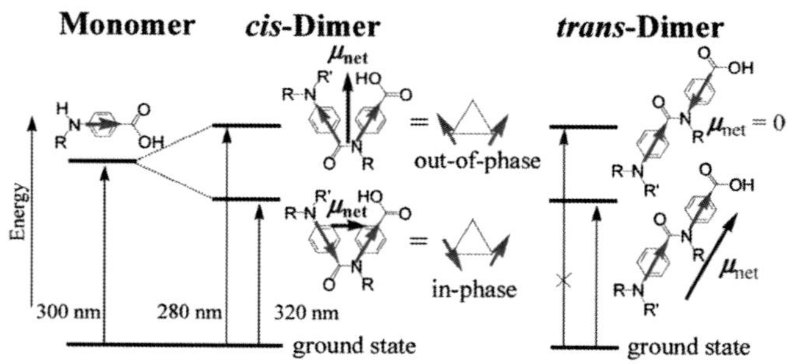

Figure 3.27 Schematic exciton band energy diagrams for *cis*- and *trans*-dimers of 4-(alkylamino)benzoic acid. The lighter and darker arrows indicate the edtms of each 4-(alkylamino)benzoyl choromophore and the net transition electric moments, respectively.[20]

arrangements of the edtms, Figure 3.27. The magnitude of the net transition moment of the higher energy state of the *cis*-dimer is predicted to be about 1.73 times larger than that of the lower energy state. The absorption intensity associated with a transition from the groundstate to the higher energy exciton state would, therefore, be expected to be more intense than that to the lower energy exciton state, as is observed in Figure 3.26. If the dimer adopts a *trans* conformation, the transition to the higher exciton state becomes optically forbidden due to a cancellation of the net moment, which provides further evidence against the presence of a *trans* dimer structure. If a cyclic planar trimer were to be formed, three exciton states would be anticipated based on the presence of a C_3 symmetry axis. The transition to the lowest energy exciton state is optically forbidden due to a cancellation of the net transition dipole moment, Figure 3.28, while, in contrast, the transition to the higher energy state, which has one nodal plane and is a doubly degenerate state, is optically allowed. The exciton band structure of a helical model compound with three chromophoric units per turn is presented in the right-hand side of Figure 3.28. By analogy with the cyclic trimer, the number of allowed electric

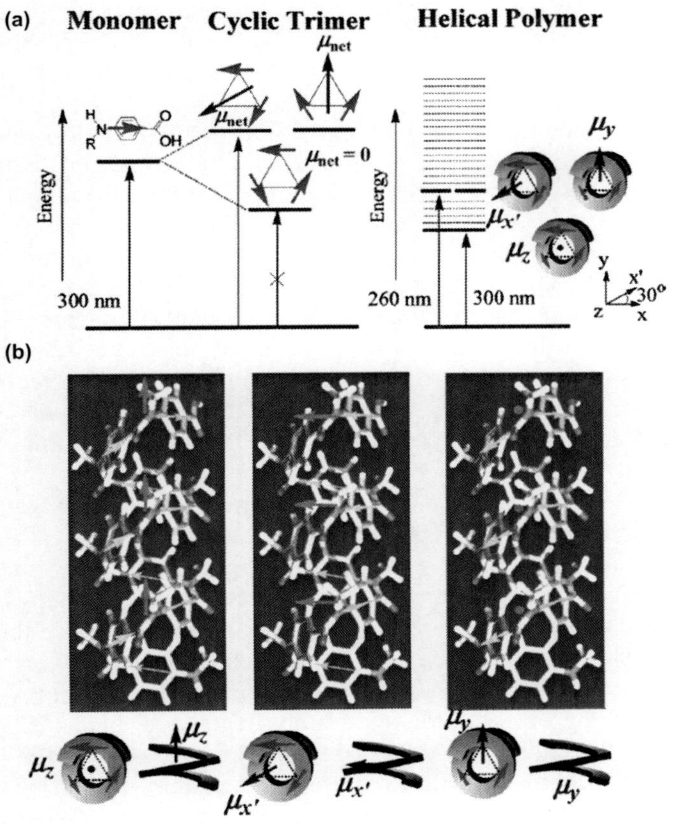

Figure 3.28 (a) Schematic exciton band energy diagrams for a cyclic trimer and a helical polymer with three light-absorbing units per turn. The smaller and larger arrows indicate the edtms of each 4-aminobenzoyl chromophore and the net transition electric moments, respectively. The dotted bands correspond to optically forbidden exciton bands for the polymer. (b) Optically allowed exciton states of the helical N-alkylated *p*-benzamide polymer. The geometry of the model oligomer was built using molecular mechanics calculations.[20]

dipole transitions of the larger polymers is severely limited for symmetry reasons, although a quasi-continuous exciton band is generated. In the context of the synthetic polyamide, one of the optically allowed exciton states corresponds to the lowest energy exciton state, in which all of the edtms couple, in a similar manner to the in-phase coupling of the *cis*-dimer and cyclic trimer. In a helical trimer this low energy exciton state gains a net edtm (μ_z) perpendicular to the plane, since each transition moment is not aligned within the same plane. The other allowed state is doubly degenerate and has one nodal plane perpendicular to the helix axis, similar to the out-of-plane coupling of the dimer and trimer. As a consequence, the absorption shoulder at around 300 nm can be readily assigned to the partially allowed transition to the lowest exciton state, which is polarised parallel to the helix axes, while the absorption band in the 260–270 nm region can be assigned to a transition to the doubly degenerate higher energy exciton state, which is polarised perpendicular to the axis. Figure 3.28 depicts the alignments of the net edtms and how they are produced.

In order to characterise the CD spectroscopy fully, the alignment of the mdtms, ***m***, also needs to be taken into consideration, since this is required to determine the sign of the rotational strengths from the Rosenfeld equation, equation (1.17). In the case of the synthetic polyamides, the two coupling modes that give rise to the lower energy positively signed and the higher energy negatively signed CD bands are associated with net edtms that lie parallel and perpendicular to the helix axis, respectively. Since the coupling between $\boldsymbol{\mu}$ and ***m*** is described as a scalar product, only the relative alignments of these transition moments with respect to the helix axis are required to be able accurately to determine the origin of the optical activity, since the alignment of ***m*** is such that when viewed from its tip, the rotation of charge is anticlockwise.

An overview of the alignment of the transition moments for right-handed ((*P*)-helix) and left-handed ((*M*)-helix) helical polymers with three light-absorbing units per turn is presented in Figure 3.29. The lower energy exciton state can be represented as having head-to-tail couplings of neighbouring transition moments, giving rise to an mdtm which is either parallel or antiparallel to the *z*-axis because of the helical rather than cyclic planar arrangement of the edtms of the three component chromophores that form each turn. On the other hand, the doubly degenerate higher exciton state has an mdtm directed along the x'- or y-axis. In the case of the (*P*) -helix, the edtm and mdtm are in an antiparallel alignment along the x'- and y-axes, resulting in a negatively signed signal, while they have a parallel alignment along the *z*-axis, resulting in a positively signed CD signal at longer wavelengths. Since the CD spectrum of polymer **1**, Figure 3.25, contains a plus-to-minus pattern in ascending energy terms, it is therefore safe to conclude that N-alkylated poly(*p*-benzamide) preferentially adopts a right-handed (*P*)-helix geometry. It should be noted that the positive and negative CD signals have the same intensity in this context due to the "sum rule", equation (1.20).

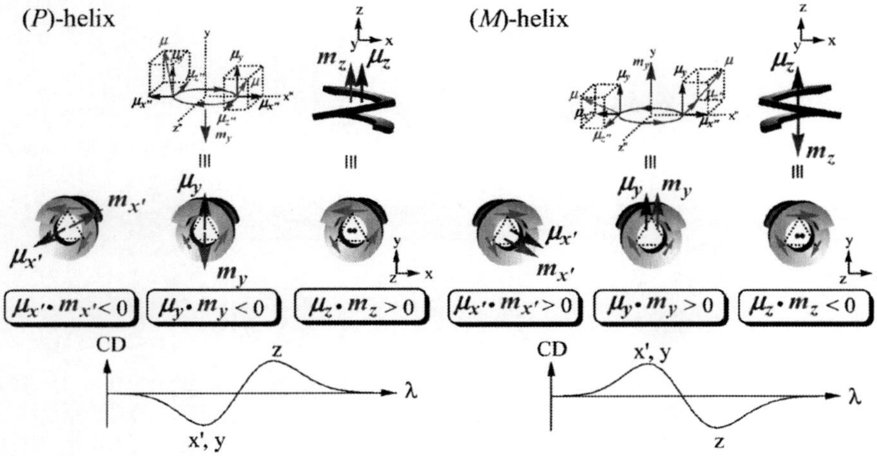

Figure 3.29 Overview of the direction of the net edtm (***μ***, dark grey) and mdtm (***m***, light grey). The theoretically predicted CD signals are shown at the bottom. Note that the x'- and z'-axes are tilted slightly, compared with the x- and z-axes. The circle around the y-axis represents the circulation of charge arising from the coupling of two transition electric moments.[20]

3.2.2 Conformation of Helical Poly(2,3-quinoxaline)s[21]

The chiral poly(2,3-quinoxaline)s, **3**, depicted in Figure 3.30 were prepared based on a block copolymerisation of 1,2-diisocyano-3,6-dimethyl-4,5-*bis*-(propoxymethyl) benzene, **2,** with a chiral pentamer, **1**. The degree of polymerisation was *ca.* 40. Conformational energy calculations for dihedral angles Ψ and Φ, Figure 3.31, predict that the energy minimum is reached at $\Psi = 135°$ when Ψ is varied in 5° steps with $\Phi = 0°$. This structure appears to be highly plausible, since X-ray structural analysis of quater(2,3-quinoxalinyl) palladium, a complex similar to **1**, demonstrated that it adopted a helical structure with $\Psi = 123–148°$.[21] The theoretical CD signal was analysed on the basis of exciton coupling theory. The absorption spectrum of quinoxaline in CH_2Cl_2 contains a weak band at 316 nm and an intense long-axis-polarised band at 233.5 nm, while the theoretical CD spectrum of a right-handed poly(2,3-quinoxaline) 20-mer with $\Psi = 135°$, Figure 3.32, exhibits a negative exciton couplet at 260 nm, see Figure 1.17. Since this pattern is the same as that observed in the spectrum of (+)-poly(2,3-quinoxaline), **3**, it is safe to assume that the quinoxaline chromophores adopt a right-handed helical arrangement.

3.3 Exciton Coupling between Non-Identical Chromophores

Although the exciton chirality method is especially useful when two identical chromophores are involved, the method can also be applied to systems

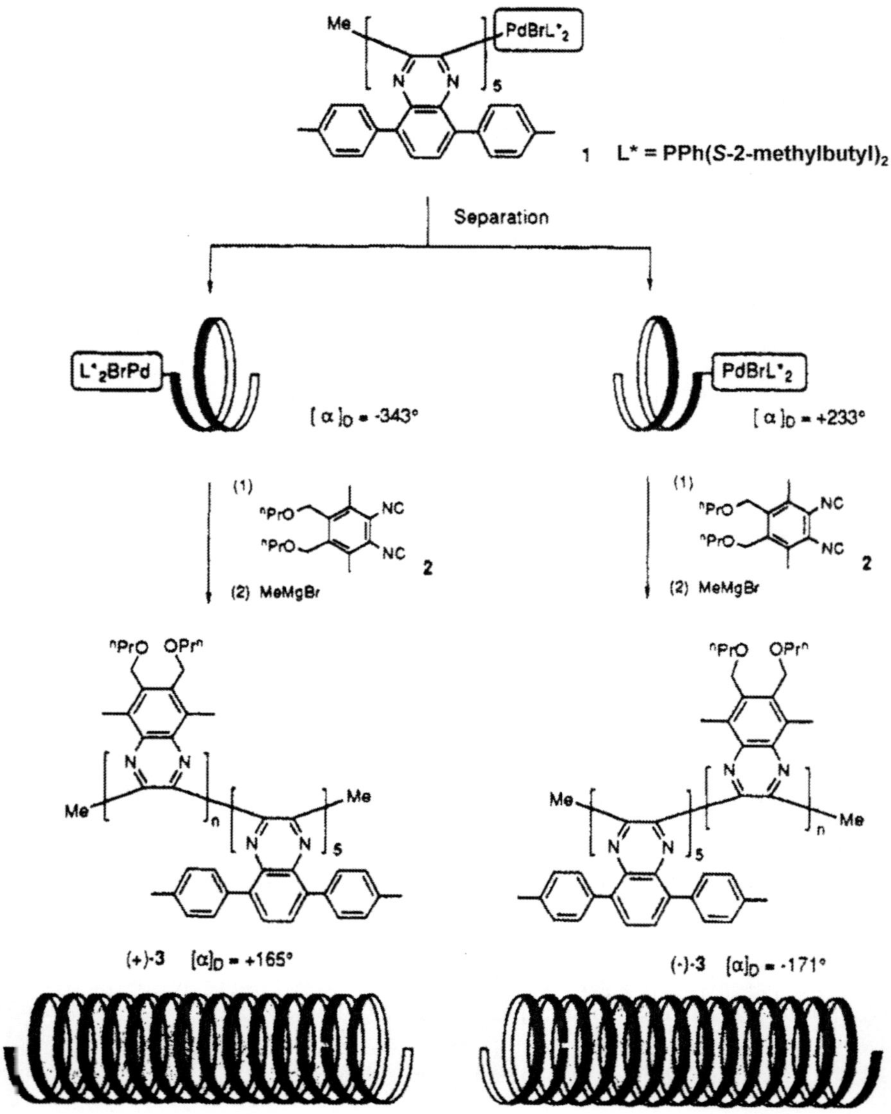

Figure 3.30 Synthetic scheme leading to poly(2,3-quinoxaline)s **3**.[21] [Reprinted with permission from *Macromolecules*, 1992, **25**, 6810. Copyright 1992, American Chemical Society.]

containing non-identical chromophores. For example, when determining the absolute configuration or conformations of natural products, it usually is not possible to introduce a chromophore identical to that already present by derivatisation. As can be readily deduced from the manner in which the exciton chirality method was applied to homodimers, there are a few

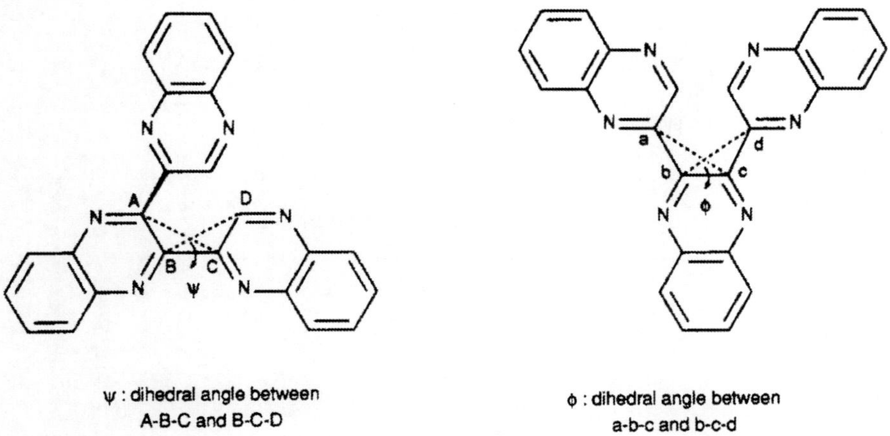

Figure 3.31 Partial structure of poly(2,3-quinoxaline)s **3**.[21] [Reprinted with permission from *Macromolecules*, 1992, **25**, 6810. Copyright 1992, American Chemical Society.]

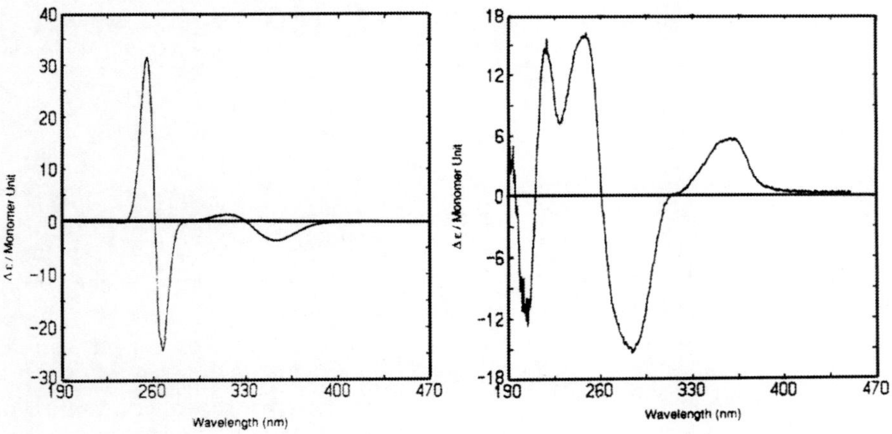

Figure 3.32 Theoretical CD curve for a right-handed poly(2,3-qunoxaline) (left) and CD spectrum of (+)-poly(2,3-quinoxaline) **3** (solvent not indicated).[21] [Reprinted with permission from *Macromolecules*, 1992, **25**, 6810. Copyright 1992, American Chemical Society.]

important features that chromophores should possess in non-identical systems. The alignment of the edtms should be well defined and both chromophores should have absorption bands at relatively close wavelengths to maximise the magnitude of the splitting of the CD band maxima. This is illustrated in Figure 3.33.[22] In the 3,4-dibenzoate structures studied, the

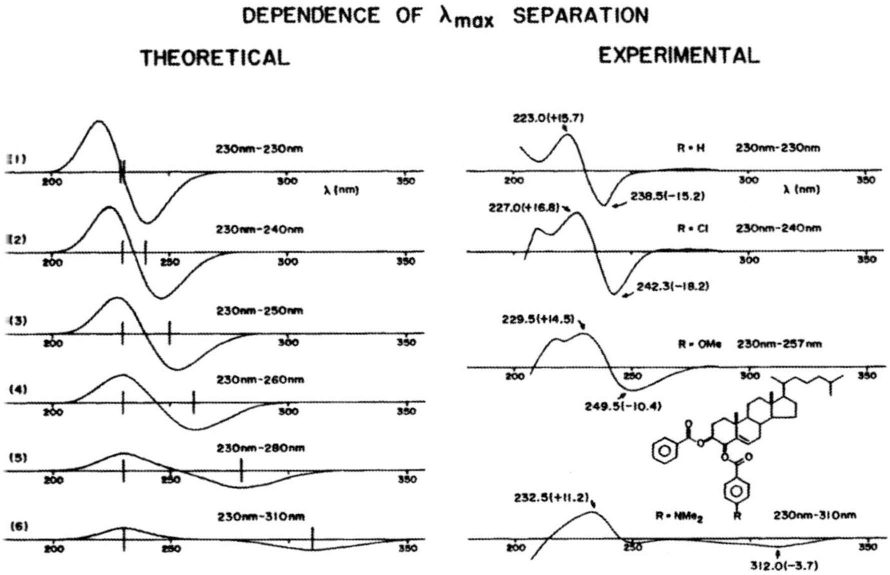

Figure 3.33 Calculated exciton-split CD bands as a function of λ_{max} separation of two different chromophores (left) and the observed CD spectra of 3-benzoate-4-*p*-substituted benzoates of cholest-5-ene-3β,4β-diols in EtOH (right).[22] [Reprinted with permission from *Pure Appl. Chem.*, 1984, **56**, 1030. Copyright 1984, IUPAC.]

absorption band maximum of the 3-OBz chromophore was kept constant while that of the 4-OBz was sequentially shifted to the red through the incorporation of a series of different *para* substituents. Excellent agreement is observed between experiment and theory. The negative sign of the lower energy band is still retained even when R = –NMe_2 and there is a band separation of 80 nm. Only the lower energy band of the exciton couplet should be considered in instances such as this because other transitions overlap in the region of the higher energy band at shorter wavelengths. If proper attention is paid so that other factors do not interfere, it is possible to determine whether there is negative or positive chirality even if only one wing of the exciton couplet can be measured in full.

3.3.1 Determination of the Absolute Configuration of Allylic Alcohols: Systems Containing a C=C Double Bond and a Benzoate[23,24]

Benzoate groups can be introduced to natural allylic alcohols to form a coupled nondegenerate system. The benzoate chromophore exhibits an allowed π → π* intramolecular charge-transfer band at 230 nm, while the

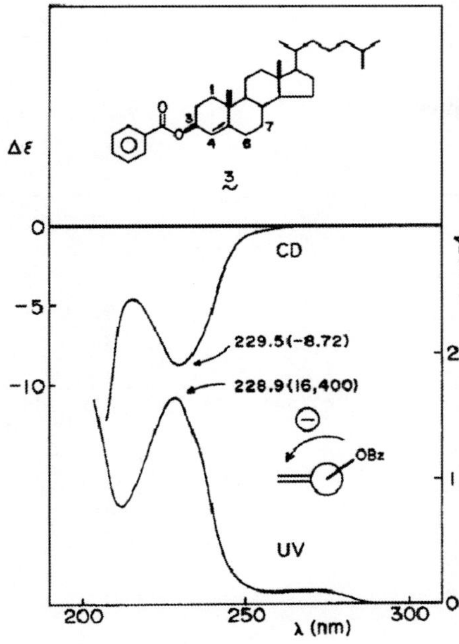

Figure 3.34 CD and absorption spectra of cholest-4-ene-3β-yl benzoate in ethanol.[23] [Reprinted with permission from *J. Am. Chem. Soc.*, 1981, **103**, 5590. Copyright 1981, American Chemical Society.]

C=C double-bond chromophore has an allowed π → π* transition at around 195 nm; both π → π* transitions are polarised along the long axes of the chromophores. If the two long axes of benzoate and double-bond chromophores constitute a positive exciton chirality, *i.e.* a right-handed helicity, the longer wavelength CD band at 230 nm, which is associated with the benzoate, is positively signed, therefore, while the second band, at shorter wavelength, at 195 nm, which is associated with a C=C π → π* transition, is negatively signed. Figure 3.34 contains the spectra of cholest-4-en-3β-yl benzoate.[23] The CD spectrum exhibits a negatively signed CD band in the region of the benzoate π → π* band at *ca.* 229 nm so the sign is consistent with a negative chirality alignment, see Figure 1.17, between the benzoate and double-bond chromophores. The configuration was, therefore, determined to be *R*. Although the negative CD signal at the instrument limit is in apparent disagreement with the theoretical prediction that the signs of the lower and higher energy CD bands should be opposite in sign, this was believed to be due in part to the presence of the B transition (~200 nm) of the benzenoid chromophore.

A similar approach has been applied to determine the absolute configuration of nemadectins, which act as potent antiparasitic antibodies against a wide range of mammalian endo– and ectoparasites.[24] Since nemadectin α

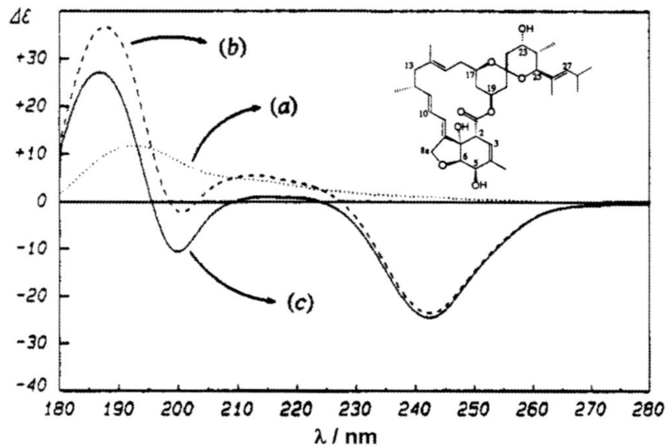

Figure 3.35 Structure of nemadectin α and CD spectra of (a) its native form, (b) 5-benzoate derivative and (c) their difference.[24] [Reprinted with permission from *J. Am. Chem. Soc.*, 1992, **114**, 7371. Copyright 1992, American Chemical Society.]

contains diene and ester chromophores, in addition to the allylic OH moiety, the CD signal of the native compound and its 5–benzoate were measured and their difference plotted, Figure 3.35. The CD band arising from the benzoate chromophore at 241 nm is negatively signed, so there is a negative chirality alignment between the C5–O bond and the 3–ene, indicating that the configuration of 5–OH is β or *R*. By measuring the area (rotational strength) of the negative CD envelope around 241 nm and comparing it to the positive one at *ca.* 186 nm, it was confirmed that these bands form an exciton couplet, since the positive and negative rotational strengths of the exciton couplet are expected to be approximately the same due to the "sum rule".

3.3.2 Application of the Exciton Chirality Method to Conjugated Enones, Esters and Lactones[25]

The examples provided in Table 3.4 demonstrate that a positive chirality between the edtms of diverse carbonyl-containing chromophores results in exciton-split CD curves having positively signed lower energy CD bands. From the data in this table, the following conclusions can be drawn. Firstly, the benzoate intramolecular charge-transfer transitions or the ^1L_a transition at 230 nm ($\varepsilon = 14\ 000$) interact with the enone $\pi \rightarrow \pi^*$ transition in the 230–260 nm region ($7000 < \varepsilon < 15\ 000$), Figure 3.36, resulting in split CD curves which can be used to determine the enone-hydroxyl chirality (entries 1–7). Whether the 205–230 nm region enone extrema are involved or not is unclear. Secondly, the

Table 3.4 Exciton-split CD data of conjugated enone, lactone and ester chromophores.[25] [Reprinted with permission from *J. Am. Chem. Soc.* 1974, **96**, 266. Copyright 1981, American Chemical Society.]

Entry	*Compound*	*Chirality*	Δε (nm) *1st band*	*2nd band*
1	3β-Hydroxycholest-5-en-7-one			
a	3-Benzoate	(+)	+7.3(237)	−18.9(217)
b	3-*p*-Chlorobenzoate	(+)	+16.7(246)	−21.6(221)
2	6β-Hydroxycholest-4-en-3-one			
a	6-Benzoate	(−)	−15.5(241)	+21.1(211)
b	6-*p*-Chlorobenzoate	(−)	−24.4(247)	+23.0(224)
c	6-*p*-Methoxybenzoate	(−)	−16.8(259)	+21.4(238)
3	6α-Hydroxycholest-4-en-3-one			
a	6-Benzoate	(+)	−13.5(238)	
b	6-*p*-Chlorobenzoate	(+)	+19.9(247)	
4	7α-Hydroxycholest-4-en-3-one-7-benzoate	(+)	+13.1(250)	−3.3(230)
5	Diosgenin derivative	(−)	−25.1	+3.2(221)
6	3β,14α-Dihydroxy-5α-cholest-7-en-6-one 3-*p*-Chlorobenzoate	(−)	−14.7(250)	+8.9(228)
7	Ponasterone A derivative	(−)	−7.4(251)	+7.7(228)
8	Shikimic acid derivative	(+)	+6.5(228)	
9	Dehydrovomifoliol	(+)	+38.4(242)	−30.2(208)
10	γ-Metasantonin	(−)	−23.0(228)	+11.3(208)

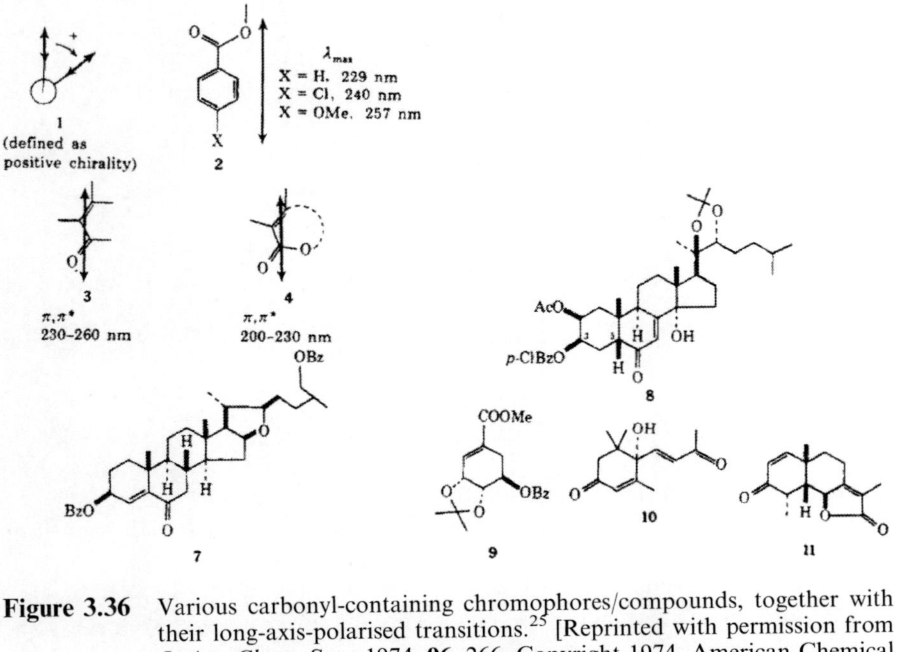

Figure 3.36 Various carbonyl-containing chromophores/compounds, together with their long-axis-polarised transitions.[25] [Reprinted with permission from *J. Am. Chem. Soc.*, 1974, **96**, 266. Copyright 1974, American Chemical Society.]

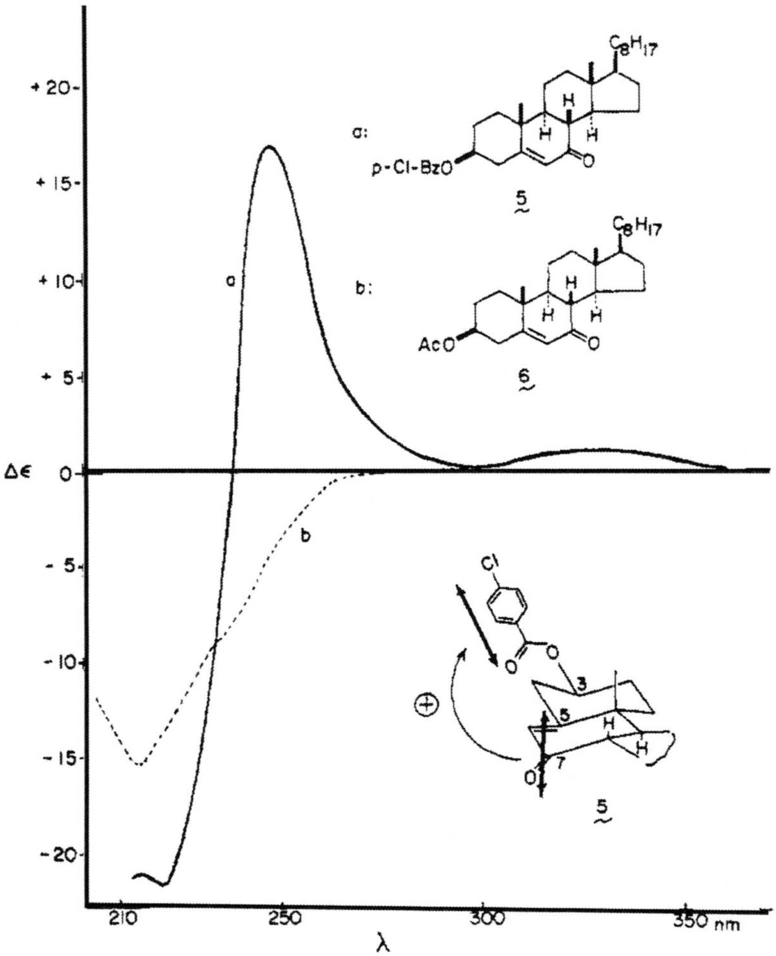

Figure 3.37 CD spectra of 3β-hydroxycholest-5-en-7-one 3-*p*-chlorobenzoate (**1**) (curve a) and 3-acetate (**2**) (curve b) in MeOH.[25] [Reprinted with permission from *J. Am. Chem. Soc.*, 1974, **96**, 266. Copyright 1974, American Chemical Society.]

differences between the exciton-split signal of the enone–benzoate, curve a in Figure 3.37, and the signal of the non-interacting enone–acetate, curve b, is obvious. Judging from the magnitude of the resulting split CD signal, it is advisable to choose a *para*-substituted benzoate with a long axis transition maximum closer to the enone $\pi \rightarrow \pi^*$ maximum (see entries 2a, b and c). Thirdly, as observed in entries 6 and 7 of Table 3.4, the enone-benzoate interaction can be identified when the systems are not homoallylic (entries 1 and 4) or allylic (entries 2, 3 and 5). In some cases, the higher energy CD band is very weak or virtually unobservable (entries 3, 5 and 8), but this does not detract from the utility of the method, since the wavelength of the lower energy CD band remains unchanged. Fourthly, the exciton chirality method can be applied to sesquiterpenoids (entry 10, γ-metasantonin).

References

1. N. Harada, S.-M. L. Chen and K. Nakanishi, *J. Am. Chem. Soc.*, 1975, **97**, 5345.
2. S. M. L. Chen, N. Harada and K. Nakanishi, *J. Am. Chem. Soc.*, 1974, **96**, 7352.
3. (a) N. Harada, V. Takuma and H. Uda, *J. Am. Chem. Soc.*, 1976, **98**, 5408; (b) N. Harada, V. Takuma and H. Uda, *J. Am. Chem. Soc.*, 1978, **100**, 4029.
4. H. Liu and K. Nakanishi, *J. Am. Chem. Soc.*, 1981, **103**, 5591.
5. H. Liu and K. Nakanishi, *J. Am. Chem. Soc.*, 1982, **104**, 1178.
6. K. Tsubaki, K. Takaishi, H. Tanaka, M. Miura and T. Kawabata, *Org. Lett.*, 2006, **8**, 2587.
7. S. Matile, N. Berova, K. Nakanishi, S. Novkova, I. Philipova and B. Blagoev, *J. Am. Chem. Soc.*, 1995, **117**, 7021.
8. S. Matile, N. Berova, K. Nakanishi, J. Fleischhauer and R. W. Woody, *J. Am. Chem. Soc.*, 1996, **118**, 5198.
9. A. Muranaka, M. Okuda, N. Kobayashi, K. Somers and A. Ceulemans, *J. Am. Chem. Soc.*, 2004, **126**, 4596.
10. M. Kasha, H. R. Rawls and M. A. El-Bayoumi, *Pure Appl. Chem.*, 1965, **11**, 371.
11. V. V. Borovkov, I. Fujii, A. Muranaka, C. A. Humbry, T. Tanaka, A. Ceulemans, N. Kobayashi and Y. Inoue, *Angew. Chem. Int. Ed.*, 2004, **43**, 5481.
12. (a) D. Gargiuro, F. Derguini, N. Berova, K. Nakanishi and N. Harada, *J. Am. Chem. Soc.*, 1991, **113**, 7046; (b) N. Berova, D. Gargido, F. Derguini, K. Nakanishi and N. Harada, *J. Am. Chem. Soc.*, 1993, **115**, 4769.
13. A. Collet and G. Gottarelli, *J. Am. Chem. Soc.*, 1981, **103**, 204.
14. C. A. Ndraud, C. Garcia and A. Collet, in *Circular Dichroism*, ed. N. Berova, K. Nakanishi and R. W. Woody, Wiley-VCH, New York, 2000, p. 383.
15. W. G. Hill, S. F. Mason and R. D. Peacock, *J. Chem. Soc., Perkin Trans. II*, 1977, 1262.
16. J. R. Platt, *J. Chem. Phys.*, 1951, **19**, 263.
17. J. Petruska, *J. Chem. Phys.*, 1961, **34**, 1120.
18. J. Sagiv, *Tetrahedron*, 1977, **33**, 2303.
19. J. Canceill, A. Collet, J. Gabard, G. Gottarelli and G. P. Spada, *J. Am. Chem. Soc.*, 1985, **107**, 1299.
20. A. Tanatani, A. Yokoyama, I. Azumaya, Y. Takakura, C. Mitsui, M. Shiro, M. Uchiyama, A. Muranaka, N. Kobayashi and T. Yokozawa, *J. Am. Chem. Soc.*, 2005, **127**, 8553.

21. Y. Ito, E. Ihara, M. Murakami and M. Sisido, *Macromolecules*, 1992, **25**, 6810.
22. K. Nakanishi, M. Kuroyanagi, H. Nambu, E. M. Oltz, R. Takeda, Gregory, L. Verdine and A. Zask, *Pure Appl. Chem.*, 1984, **56**, 1030.
23. N. Harada, J. Iwabuchi, Y. Yokota and H. Uda, *J. Am. Chem. Soc.*, 1981, **103**, 5590.
24. L.-C. Lo, N. Berova, K. Nakanishi, G. Schlingmann, G. T. Carter and D. B. Boders, *J. Am. Chem. Soc.*, 1992, **114**, 7371.
25. M. Koreeda, N. Harada and K. Nakanishi, *J. Am. Chem. Soc.*, 1974, **96**, 266.
26. N. Kobayashi and A. Muranaka, *J. Synth. Org. Chem., Jpn.*, 2006, **64**, 735.

CHAPTER 4

Cyclodextrin Inclusion Compounds

Cyclodextrins (CyDs) are 1,4-linked oligomers of D(+)-glucose formed during the degradation of starch by an amylase of *Bacillus macerans*.[1] Those comprised of 6, 7 and 8 glucose residues are referred to as α-, β- and γ-CyD, respectively, Figure 4.1. It has long been known that CyDs can trap aromatic compounds, alkyl halides and gas molecules as guest molecules in their central cavities, resulting in the formation of inclusion complexes,[2] and that a CD signal can be induced when a suitable chromophore is involved in this regard. The orientation of the chromophore can be elucidated by analysing the CD spectra, or alternatively, if the inclusion type has been established, the alignment of the edtm can be determined.

There are several factors that need to be taken into consideration when studying the CD spectroscopy of CyDs:

1. Given that water is generally used as the solvent, measurements can be made at wavelengths as short as 180 nm.
2. The formation of an inclusion complex is generally instantaneous so spectra can be measured immediately after mixing the chromophore with the cyclodextrin.
3. Although α-, β- or γ-CyD can be selected depending on the size of the chromophore, chromophores larger than the cavity of γ-CyD cannot be trapped. In most cases, β-CyD has been used.
4. The formation of a product with a CyD : guest ratio of 1 : 1 is generally a prerequisite, although chirality information can sometimes also be derived from inclusion compounds with 2 : 1 or 1 : 2 ratios.

Circular Dichroism and Magnetic Circular Dichroism Spectroscopy for Organic Chemists
By Nagao Kobayashi, Atsuya Muranaka and John Mack

Published by the Royal Society of Chemistry, www.rsc.org

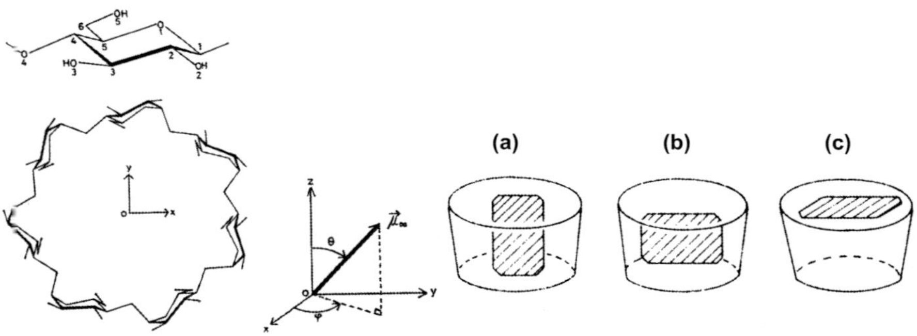

Figure 4.1 (a) The structure of β-CyD. The structures of α-CyD and γ-CyD contain one unit less or more of this cylindrical structure.[5] (b) Coordinate system of the electric dipole moment.[5] (c) Axial (left), equatorial (centre) and lid-type inclusion.[10] [Reprinted with permission from *Bull. Chem. Soc. Jpn.*, 1975, **48**, 375. Copyright 1975, The Chemical Society of Japan.]

4.1 Induced Circular Dichroism of CyDs

Given that CyDs are composed of chiral glucose units, the CD signal is induced in the absorption bands of the achiral guest molecules. The rotational strength values, R_{0a}, of intense electric dipole allowed bands interacting with far-UV transitions of vicinal groups, such as those associated with $\sigma \rightarrow \sigma^*$ bands, can be determined using the coupled oscillator model of Kirkwood,[3] see equations (1.22) and (1.23), as applied by Tinoco:[4]

$$R_{0aj} = \pi \nu_a \nu_0{}^2 \boldsymbol{\mu}_{0a}{}^2 (\alpha_{33} - \alpha_{11})_j (\mathrm{GF})_j / \{\mathrm{h}(\nu_0{}^2 - \nu_a{}^2)\} \tag{4.1}$$

$$(\mathrm{GF})_j = (1/\mathbf{r}_j{}^3)\left[\mathbf{e}_{0a} \cdot \mathbf{e}_j - 3(\mathbf{e}_{0a} \cdot \mathbf{r}_j)(\mathbf{e}_j \cdot \mathbf{r}_j)/\mathbf{r}_j{}^2\right]\mathbf{e}_{0a}(\mathbf{e}_j \cdot \mathbf{r}_j) \tag{4.2}$$

In this context, $\mathbf{e}_j$ is the unit vector in the direction of the main symmetry axis of the CyD; $\mathbf{e}_{0a}$ is the unit vector in the direction of the electric dipole moment ($\boldsymbol{u}_{0a}$) of the transition from the groundstate (0) to the excited state (a) in the guest molecule; ν_a is its wavenumber; α_{33} and α_{11} are bond polarisabilities at zero frequency parallel and perpendicular, respectively, to the main symmetry axis of the CyD; $\mathbf{r}_j$ is the vector pointing from the centre of the guest molecule to each bond in CyD; ν_0 is the averaged wavenumber of the electronic transitions of the CyD. The origin of the coordinates is generally placed at the centre of the CyD ring and the alignment of $\mathbf{e}_{0a}$ is expressed using spherical polar coordinates. θ is the angle of $\mathbf{e}_{0a}$ with respect to the z-axis, the main symmetry axis of CyD, and ϕ is the angle which the projection of $\mathbf{e}_{0a}$ into the x/y plane forms relative to the x-axis. It was ultimately found that R_{0a} does not depend on the values of ϕ, β but is determined instead by the value

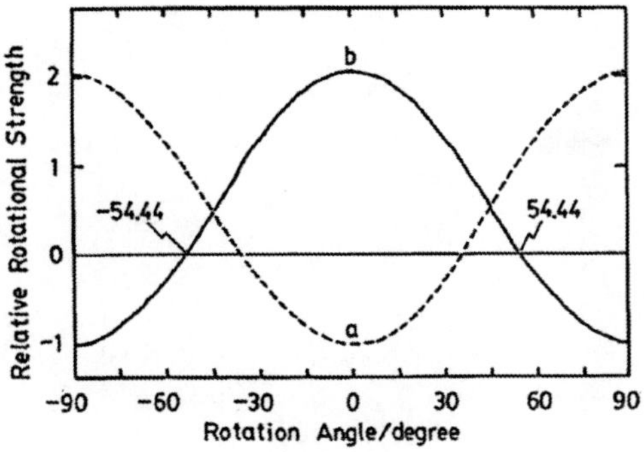

Figure 4.2 Dependence of the calculated rotational strength of a CyD-chromophore complex on the rotation angle in the *xz*-plane. Curves **a** and **b** are the transitions normal and parallel to the molecular axis of the CyD, respectively.[17]

of θ. Since $\mathbf{e}_{0a} = (\sin\theta, 0, \cos\theta)$, equation (4.1) can be expressed in the following format:[5]

$$R_{0a} = [A_{0a} + B_{0a}\cos 2\theta + C_{0a}\sin 2\theta]\boldsymbol{\mu}_{0a}{}^2 \qquad (4.3)$$

the magnitude of C_{0a} was found to be a factor of 1000 smaller than that of B_{0a}, so the presence of $B_{0a}\cos 2\theta$ in equation (4.3) effectively indicates that R_{0a} takes its maximum value at $\theta = 0°$and its minimum value at $\theta = 90°$. The detailed dependence of R_{0a} on θ is shown in Figure 4.2. As can be judged from curve b in this figure, electronic transitions whose directions are aligned at an angle lower than 54.44° from the axis of the CyDs produce a positive ICD signal, while those with an angle larger than 54.44° yield a negative ICD signal. The ICD intensity at $\theta = 90°$ is only half that at $\theta = 0°$. It can also be demonstrated that the sign of the calculated rotational strength does not change as a result of the movement of guest molecules along the *x*-, *y*- and *z*-axes, although the magnitude is affected.[6] It is possible, therefore, to determine the polarisation direction of the absorption band of chromophore molecules from the sign of the ICD spectra of the 1 : 1 chromophore : CyD complex. Conversely, the orientation of guest molecules trapped in the CyD cavity can be inferred from the sign of the ICD spectra, if the polarisations of the bands of the guest molecules are known.

4.1.1 Inclusion of Pyrene[7,8] in β-CyD at Room Temperature and in γ-CyD at 70°

Figure 4.3 contains the absorption and ICD spectra of pyrene in β-CyD. The wavelengths of the peaks and troughs in the ICD spectrum lie close to the

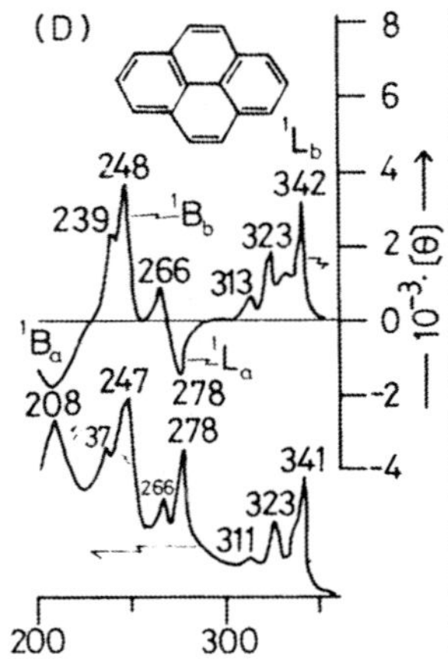

Figure 4.3 Electronic absorption and CD spectra of pyrene trapped in the cavity of β-CyD in water.[7]

band centres observed in the electronic absorption spectrum, despite the complexity that is introduced by the vibrational nature of these bands. Since an "*axial inclusion*" in which the long axis of the guest molecule is parallel to the molecular axis of the CyDs is the only conceivable inclusion type for pyrene, the negatively signed ICD bands at 278 and 208 nm can only be assigned to the short-axis-polarised transitions, so the other positively signed ICD bands are polarised along the long axis of the molecule. The assignments at wavelengths longer than 270 nm were in complete accord with those obtained using a cholesteric mesophase.[9] Although two pyrene molecules are trapped in γ-CyD at room temperature and exhibit a left-handed chirality, at 70 °C the spectra became similar to those in Figure 4.2 owing to the loss of one of the pyrene molecules.[8]

4.1.2 Inclusion of 2,3-Diaminonaphthalene and 1,8-Diaminonaphthalene in β-CyD[5]

Figure 4.4 contains the UV region of the absorption and induced circular dichroism (ICD) spectra of 2,3-diaminonaphthalene and 1,8-diamino-naphthalene in β-CyD. A consideration of the size and shape of the guest molecule makes it clear that axial inclusion is the only possible structure for 2,3-diaminonaphthalene. The CD spectrum, therefore, contains a positively signed band at *ca.* 240 nm due to the long-axis-polarised ^1B_b transition.[7] In marked contrast, the ICD spectrum of 1,8-diaminonaphthalene contains a negatively signed ^1B_b band, so the structure was judged to be an example of an unusual "*equatorial inclusion*" with the short axis lying parallel to the molecular axis of β-CyD.[5]

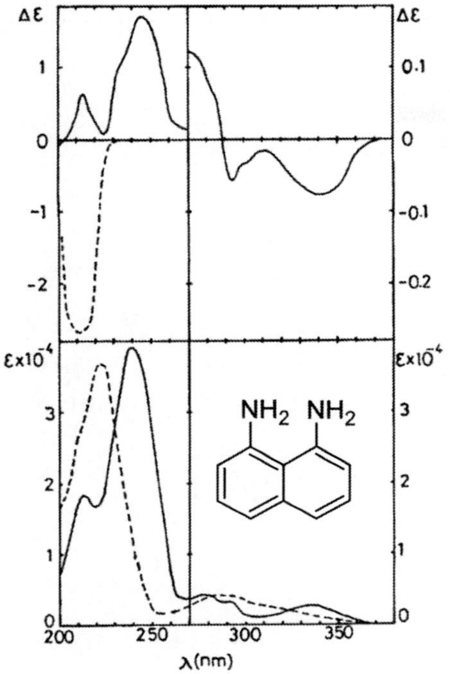

Figure 4.4 Electric absorption and CD spectra of 1,8-diaminonaphthalene (broken lines) and 2,3-diaminonaphthalene (solid lines) in the presence of β-CyD in water.[5] [Reprinted with permission from *Bull. Chem. Soc. Jpn.*, 1975, **48**, 375. Copyright 1975, The Chemical Society of Japan.]

4.1.3 Lid-type Inclusion of Pyrene-1,3,6,8-tetrasulphonate Anion by β-CyD[10]

In order to determine the orientation of guest molecules in CyDs, the band polarisations of the chromophore must have been explicitly elucidated. A pyrene derivative, pyrene-1,3,6,8-tetrasulphonate anion, see Figure 4.2, was selected from this standpoint to investigate a conjectured new type of inclusion. In this molecule four sulphonate groups are symmetrically linked to the pyrene π-system and this was believed to prevent axial inclusion for steric reasons. Since the inside of the CyD cavity is hydrophobic, it was believed that the more hydrophobic pyrene structure would be included first rather than a sulphonate group, enabling the formation of an ideal "*lid-type*" inclusion. After the 1 : 1 stoichiometry of the complex was confirmed by Job's continuous variation experiments, this inclusion structure was confirmed as shown in Figure 4.5. In this figure, only the band at 283 nm is polarised along the short axis, while all the others are polarised along the long axis. Negatively signed bands are observed in the CD spectrum in this figure corresponding to all of the absorption bands, indicating that both the short- and long-axis-polarised transitions are normal to the molecular axis of β-CyD. Since these phenomena would only

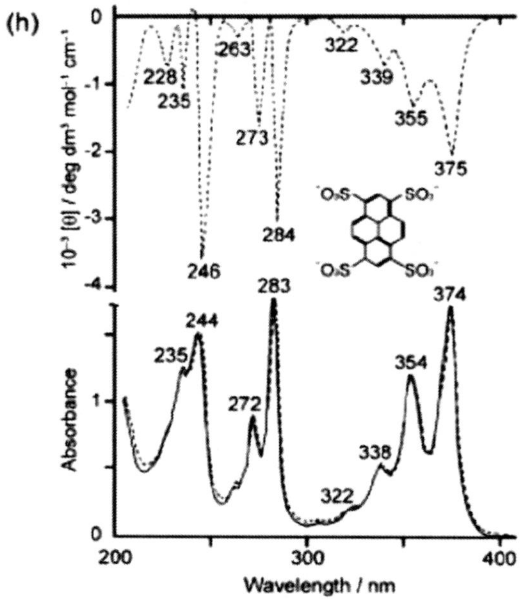

Figure 4.5 Electric absorption and CD spectra of 1,3,6,8-pyrenetetrasulphonate in the presence of β-CyD in water[10] at pH = 9.0.

be anticipated for lid-type inclusions, this was judged to be the first example of a "lid-type" inclusion.

4.1.4 Inclusion of 4,4'-Substituted Biphenyls by β-CyD[7]

One of the most characteristic aspects of the optical properties of biphenyl and its derivatives is the appearance of an intense, structureless envelope of absorption intensity in the 260–280 nm region, which is referred to as the conjugation band. This band is known to be sensitive to variations in the torsional angle around the central C–C bond of biphenyl, which results in changes in both the location of the band maximum and its intensity, due to differences in the extent of π-conjugation in the two benzene rings.[11] An MCD spectroscopy study of these compounds demonstrated that the band assignments in this spectral region can vary based on the magnitude of the torsional angle.[12] Figure 4.6 contains the spectra in the presence of β-CyD. All of the compounds must adopt an axial inclusion structure based on their molecular shape and exhibit ICD curves with a change in sign from minus to plus in ascending energy terms, corresponding to the structureless absorption band described above, indicating the presence of at least one short- and one long-axis-polarised transition in this region of the spectrum.[13] This was further confirmed using a thermotropic cholesteric mesophase formed from a mixture of cholesteryl nonanoate and cholesteryl chloride (weight ratio = 7 : 3). This mixture forms liquid crystals at temperatures ranging from 70 to 18 °C when

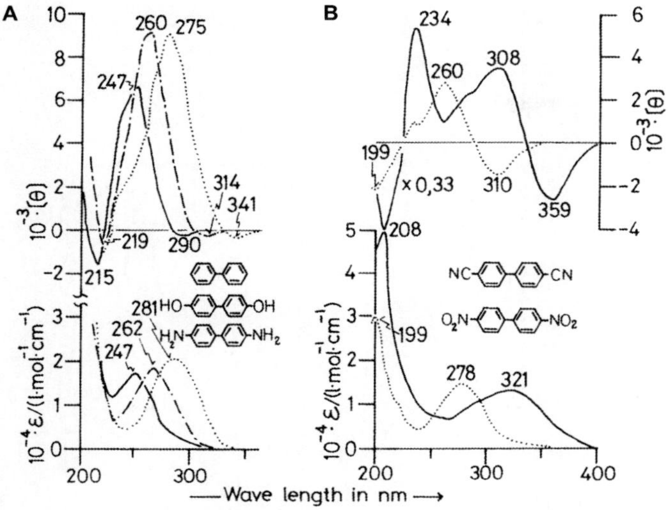

Figure 4.6 Electronic absorption and CD spectra of 4,4'-substituted biphenyls trapped in the cavity of β-CyD in water.[7]

allowed to cool after heating. The helical sense of this cholesteric phase is left-handed. The substrate in this phase shows positive and negative ICD corresponding to the short- and long-axis-polarised transitions. Below 230 nm, biphenyl, 4,4'-biphenyl diol, 4,4'-dinitrobiphenyl and 4,4'-biphenylene dicyanide exhibit troughs at 215, 219, 208 and 199 nm respectively, which may be associated with the theoretically predicted short-axis-polarised transitions.[12]

4.1.5 Inclusion of Bipyridyls by β-CyD[7]

The spectra for 4,4'- and 2,2'-bipyridyl are shown in Figure 4.7. As anticipated based on the results of theoretical calculations,[14] the ICD spectrum of 4,4'-bipyridyl contains negatively and positively signed bands in the 220–300 nm region, while in contrast, two positively signed ICD bands are observed in the spectrum of 2,2'-bipyridyl. Thus, in the 4,4'-bipyridyl spectrum, the two bands in the lower energy region are short- and long-axis polarised, while in the 2,2'-bipyridyl spectrum these bands are both long-axis polarised. The sign of the ICD band at 200 nm is negative in the former case and positive in the latter. This is consistent with the theoretical prediction[14] that 4,4'-bipyridyl has a short-axis-polarised transition at 198 nm, while 2,2'-bipyridyl has a long-axis-polarised transition at 195 nm.

4.1.6 Disposition of Ferrocene in β- or γ-CyD[15]

Theoretical calculations[16] predicted that ferrocene (Fc) can bind to β-CyD in both axial and equatorial orientations, since the energy difference between these

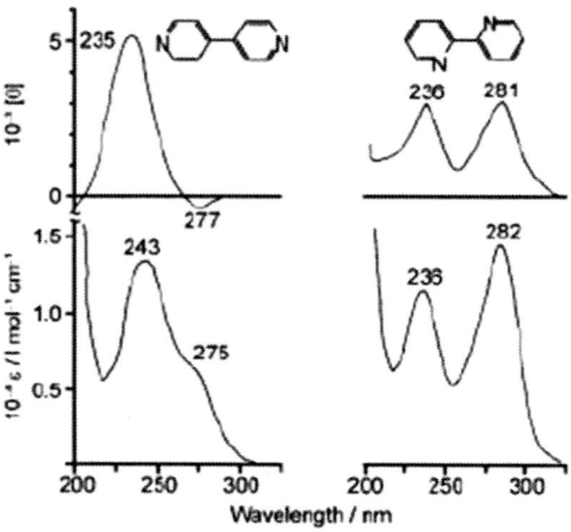

Figure 4.7 Electronic absorption and CD spectra of 4,4'- and 2,2'-bipyridyl trapped in the cavity of β-CyD in water.[7]

structures is too small to be significant. There has been uncertainty over the band polarisation directions in Fcs.[17] As a result, despite the publication of over 20 articles on the CyD complexes of Fc derivatives, the orientation of the Fcs in the CyD cavity could not be definitively determined. By preparing an Fc molecule, which can bind to CyD cavities in only one binding mode, it is possible to infer the orientation of the other Fc derivatives by comparing their ICD spectra. It may also be possible to predict the polarisation directions of the absorption bands of Fc derivatives. An Fc molecule with two cyclopentadienyl rings linked by a crown ether unit was prepared and the ICD spectrum was recorded in the presence of β-CyD, Figure 4.8. In this case, only an equatorial inclusion is possible. The absorption spectrum is similar to that of Fc, Figure 4.8, so the attached crown ether unit clearly does not have a significant effect on the electronic structure of the Fc chromophore. Negatively and positively signed ICD peaks are observed at *ca.* 430 and 220 nm corresponding to the main absorption bands. It is safe to conclude, therefore, that equatorial inclusion produces a negatively signed ICD band in the 400–550 nm region. Since unsubstituted Fc exhibits negative and positive ICD in this region in the presence of γ-CyD and β-CyD, respectively, Figure 4.8, Fc-γ-CyD and Fc-β-CyD complexes were determined to adopt equatorial and axial inclusion structures. It has been proposed that the absorption band in the longest wavelength region of the Fc spectrum contains two components with maxima between 420–430 nm and 460–470 nm. Both of these bands have been assigned to $d \rightarrow d$ transitions.[17] Interestingly, an ICD peak in the axially included Fc-β-CyD system and ICD troughs in the equatorial Fc-γ-CyD system always appear at around 460 – 470 nm. As a result, the lower energy band is believed to be polarised along the Cp-Fe-Cp axis. This result is consistent with low

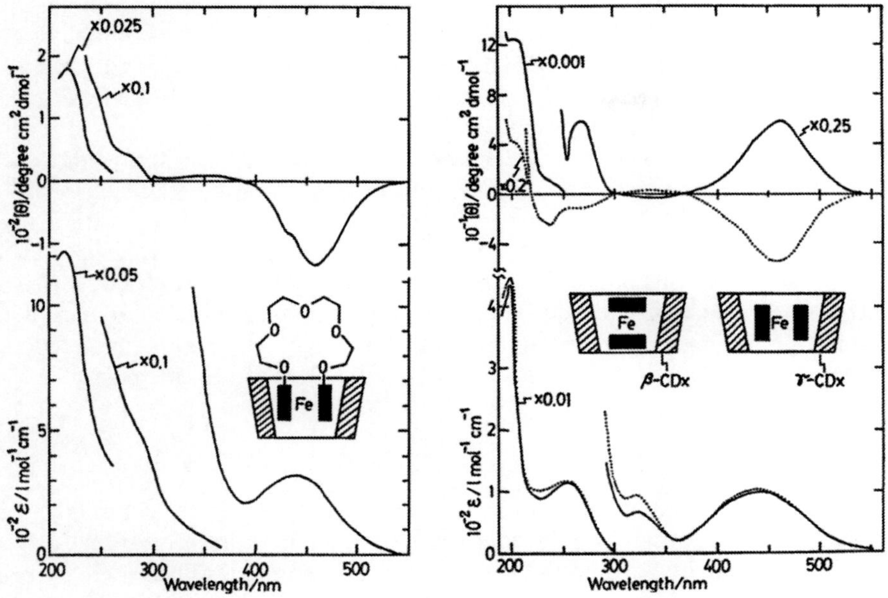

Figure 4.8 Electronic absorption and CD spectra of an Fc molecule with two cyclopentadienyl rings linked by a crown ether unit (left) trapped in the cavity of β-CyD in water. The spectra of Fc without substituents in the presence of γ-CyD and β-CyD (right) exhibits negatively and positively signed ICD in *ca.* 360–550 nm.[15]

temperature studies which indicate that the band is a symmetry forbidden $d \rightarrow d$ transition containing charge-transfer character.

Since the orientation of the Fc molecule in β-CyD and γ-CyD was clarified on this basis, the assignment of the intense band at around 200 nm could be re-examined, a subject which had long been disputed in the literature. Some groups assigned this band as a $\pi \rightarrow \pi^*$ transition of the Cp rings,[18] while others[19,20] interpreted it as a CT band between the Cp rings and the Fe(II) ion. As shown in Figure 4.8, a positively signed ICD band is observed corresponding to this absorption band for both the axial and the equatorial inclusion of Fc. If this band were associated with a $\pi \rightarrow \pi^*$ transition, the sign of the ICD band should be positive for an equatorial inclusion and negative for an axial inclusion. It should be noted that an equatorial inclusion of Fc is effectively an axial inclusion in the context of the Cp rings. The results are, therefore, not consistent with an assignment as a "pure" $\pi \rightarrow \pi^*$ transition. The question then is whether the observed ICD can be rationalised as a CT transition? The answer is that it is possible. As depicted in Figure 4.2, transitions whose polarisation directions are aligned less than 54.44° from the molecular axis of the CyDs produce a positively signed ICD band when Fc is in the CyD cavity. The d_{xz} and d_{yz} orbitals of the Fe(II) ion are aligned at 45° from the Cp-Fe-Cp axis of Fc. Hence, these orbitals will still be aligned at 45° from the molecular axis of the CyDs regardless of whether there is an axial or equatorial inclusion of the Fcs. This means that if these orbitals are involved in the CT transition between the Cp rings and the central Fe(II) ion, the sign of the ICD band becomes positive. The intense band of Fcs at around 200 nm can, therefore, be assigned definitively to a CT transition on this basis.

The polarisation directions of the other bands have also been analysed. It has been proposed, based on studies of crystals of Fcs,[21] that the bands at *ca.* 260 and 320 nm are polarised along the (Cp-Fe-Cp) *z*-axis mixed with some polarisation along the *x*- or *y*-axis. Figure 4.8 indicates that the ICD spectrum is consistent with this proposal in the context of the 260 nm band, but the sign of the ICD intensity in the 300–370 nm region suggests that the band at *ca.* 320 nm is more strongly polarised along the *x*- or *y*-directions than the *z*-direction. Otherwise, negative ICD intensity would not be observed in this region.

4.1.7 Chirality of Two Pyrene Molecules in γ-CyD[8]

The longest wavelength band of pyrene is polarised along the long axis, see Figure 4.3. The formation of a complex with a CyD : guest ratio of 1 : 2 has been confirmed by an analysis of excimer fluorescence. When two pyrene molecules are trapped in the cavity of γ-CyD a minus-to-plus pattern in ascending energy is observed, Figure 4.9. This exciton couplet signal is consistent with the presence of two pyrene molecules aligned in a left-handed conformer. In a similar manner, two 2,3-anthracenedicarboxylates were found

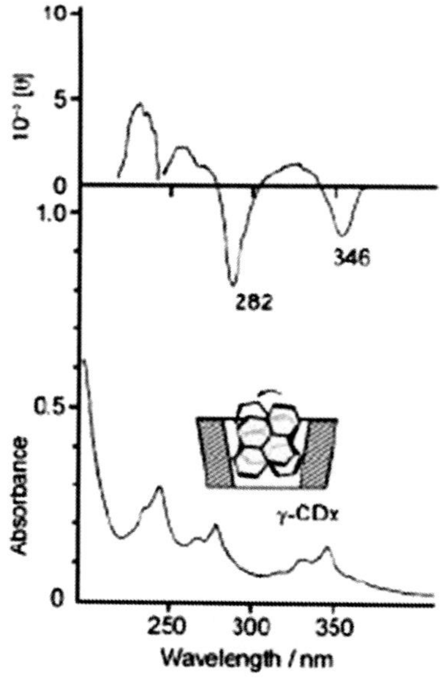

Figure 4.9 Electronic absorption and CD spectra of two pyrene molecules trapped in the cavity of γ-CyD in water.[8]

to form a right-handed conformer in γ-CyD,[22] since the ICD signal corresponding to the long-axis-polarised transition of anthracene at *ca.* 260–270 nm has a plus-to-minus sequence in ascending energy terms. In both cases, the ICD intensity results primarily from exciton coupling between the two chromophores in the CyD cavity.

References

1. F. Schardinger, *Zentr. Bakt. Parasitenk. II*, 1911, **29**, 188.
2. F. Cramer, W. Saenger and H.-Ch. Spatz, *J. Am. Chem. Soc.*, 1967, **89**, 14.
3. J. P. Kirkwood, *J. Chem. Phys.*, 1937, **5**, 479.
4. I. Tinoco, Jr., *Adv. Chem. Phys.*, 1962, **4**, 113.
5. K. Harata and H. Uedaira, *Bull. Chem. Soc. Jpn*, 1975, **48**, 375.
6. H. Yamaguchi, K. Ninomiya and R. West, *J. Chem. Soc., Faraday Trans. 2*, 1982, 78, 481.
7. N. Kobayashi, S. Minato and T. Osa, *Makromol. Chem.*, 1983, **184**, 2123.
8. N. Kobayashi, R. Saito, H. Hino, Y. Hino, A. Ueno and T. Osa, *J. Chem. Soc., Perkin Trans. II*, 1983, 1031.
9. F. D. Saeva, P. E. Sharpe and G. R. Olin, *J. Am. Chem. Soc.*, 1973, **95**, 7656.
10. N. Kobayashi, *J. Chem. Soc., Chem. Commun.*, 1988, 918.
11. H. H. Jaffe and M. Orchen, *Theory and Application of Ultraviolet Spectroscopy*, Wiley, New York, 1966, p. 384.
12. H. Uchimura, A. Tajiri and M. Hatano, *Bull. Chem. Soc. Jpn*, 1981, **54**, 3279.
13. (a) Y. Sato and M. Hatano, *Makromol. Chem.*, 1982, **183**, 971; (b) Y. Sato, A. Tajiri and M. Hatano, *Makromol. Chem.*, 1982, **183**, 989.
14. H. Uchimura, *Ph.D. Thesis*, Tohoku University, 1980.
15. N. Kobayashi, *J. Chem. Soc., Chem. Commun.*, 1990, 477.
16. F. M. Menger and M. J. Scherrod, *J. Am. Chem. Soc.*, 1988, **110**, 8606.
17. N. Kobayashi and T. Osa, *Bull. Chem. Soc. Jpn*, 1991, **64**, 1878.
18. D. R. Scott and R. S. Becker, *J. Chem. Phys.*, 1961, **35**, 516.
19. A. T. Armstrong, D. G. Carroll and S. P. McGlynn, *J. Chem. Phys.*, 1967, **47**, 1104.
20. Y. S. Sohn, D. N. Hendrickson and H. B. Gray, *J. Am. Chem. Soc.*, 1971, **93**, 3603.
21. (a) S. Yamada, A. Nakahara and R. Tsuchida, *J. Chem. Phys.*, 1954, **22**, 1620; (b) S. Yamada, A. Nakahara and R. Tsuchida, *Bull. Chem. Soc. Jpn*, 1955, **28**, 465.
22. M. Opallo, N. Kobayashi and T. Osa, *J. Incl. Phenom.*, 1989, **6**, 413.

CHAPTER 5

Metal Complexes

The CD spectroscopy of metal complexes can be divided into two groups, those with chromophore-containing ligands, such as *o*-phenanthroline and 2,2'-pyridine, and those without, such as ethylenediamine (en). The CD intensity in the former case is usually relatively strong and is often generated by exciton coupling between neighbouring chromophores, while in contrast that of the latter group tends to be relatively weak, since it is associated primarily with the electric dipole forbidden but magnetic dipole allowed $d \rightarrow d$ transitions of the central metal.

5.1 Systems which can be Analysed by Excition Coupling Theory: Absolute Configurations of Metal Complexes Containing *o*-Phenanthroline, 2,2'-Bipyridyl or Acetylacetonate

The intense CD spectra of octahedral compounds containing chromophores such as the bidentate *o*-phenanthroline (phen), 2,2'-bipyridyl (bipy), or acetylacetonate (acac) ligands, Figure 5.1, are known to arise from exciton coupling.[1] The $\pi \rightarrow \pi^*$ transitions of phen, bipy and acac are polarised in the molecular plane and are directed along either the long (x-) axis or short (z-) axis of the ligand. Upon coordination, these transitions remain largely unmodified. The CD spectra of compounds containing only one of these chromophoric ligands, for example in $[Co(en)_2phen]^3$, contain relatively weak signals due to the $\pi \rightarrow \pi^*$ bands of the ligand.[2] When two or more of these conjugated ligands are coordinated, the CD intensity associated with the long-axis-polarised transitions becomes exceedingly strong, while the intensity resulting from the short-axis-polarised transitions remains relatively weak. These observations suggest that the CD signal is produced by electrostatic coupling of edtms essentially localised on the ligands. In octahedral complexes, the short-axis-polarised

Circular Dichroism and Magnetic Circular Dichroism Spectroscopy for Organic Chemists
By Nagao Kobayashi, Atsuya Muranaka and John Mack

Published by the Royal Society of Chemistry, www.rsc.org

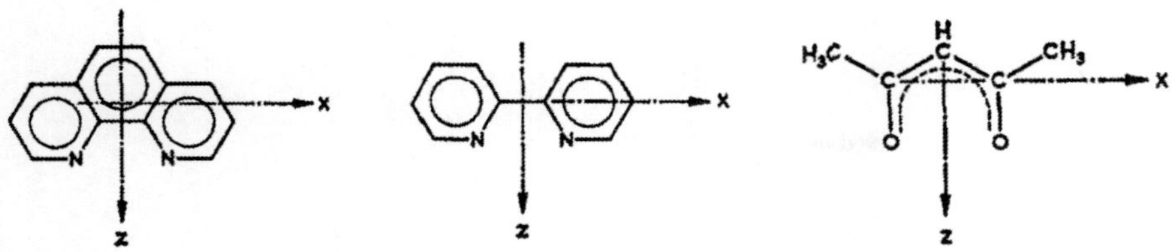

Figure 5.1 *o*-phenanthroline and 2,2'-bipyridyl molecules and acetylacetonate ion showing the long (*x*-) and short (*z*-) axes.[1] [Reprinted with permission from *Acc. Chem. Res.*, 1968, **2**, 266. Copyright 1968, American Chemical Society.]

transitions couple to form edtms, Figure 5.2, but no exciton optical activity is observed, since there is no net circulation of charge to generate a mdtm. The weak CD signal observed in the short-axis-polarised band region results from a higher order of dissymmetric perturbation, such as minor deviations from O_h symmetry. The long-axis-polarised transitions, however, couple to produce parallel edtm and mdtm vectors, so the CD intensity is strong in this region.

On the basis of MO calculations, polarised crystal spectra and CD spectral data, the lowest energy long-axis-polarised $\pi \rightarrow \pi^*$ bands of the *o*-phenanthroline and 2,2'-bipyridyl molecules and the acetylacetonate ions are known to occur at *ca.* 265, 290 and 286 nm, respectively.[1] The metal complexes containing these bidentate ligands can be grouped into five categories:

1. Identical *tris*-bidentate complexes, such as $[M(phen)_3]^{n+}$.
2. Identical *cis bis*-bidentate complexes, such as $[M(phen)_2X_2]^{n+}$.
3. Non-identical *cis bis*-bidentate complexes, such as *cis*-$[M(phen)(bipy)X_2]^{n+}$.
4. *tris*-Bidentate complexes with two different ligands, such as $[M(phen)_2bipy]^{n+}$ and $[M(bipy)phen]^{n+}$.
5. *tris*-Bidentate complexes with three different ligands.

We will primarily consider category 1, because this represents the easiest system for learning how to analyse the CD spectroscopy of O_h symmetry complexes and the principles described in that context can also be readily applied to the other four categories. $[M(phen)_3]^{n+}$ is selected as a representative example of complexes containing three idential ligands. The long-axis-polarised $\pi \rightarrow \pi^*$ transition of the phen ligand results in an oscillating displacement of the π electrons along the *x*-axis, but this transition does not produce CD intensity by itself, since no circulation of charge is involved. When three phen molecules are coordinated octahedrally to a metal atom, a photon can cause all three ligands to become excited, if the adjacent edtms can interact. This means that the ligands are no longer independent of each other. The wavefunctions of this system can be written as simple product functions, provided there is no charge-transfer between the three phen ligands. Thus, the groundstate function

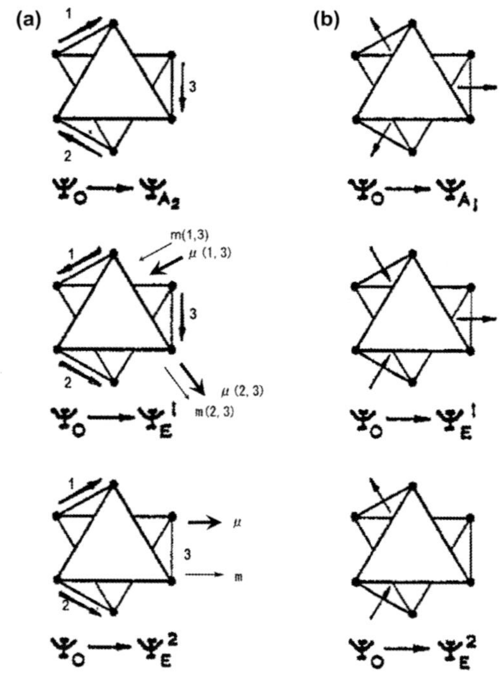

Figure 5.2 Diagrammatic representation of the three coupling modes of the long-axis-polarised ligand transitions (a) and the short-axis-polarised ligand transitions (b). In two cases of (a), the directions of electronic transition (***μ***) and magnetic transition (***m***) moments produced by coupling of two long-axis-polarised $\pi \rightarrow \pi^*$ transitions are also shown in the parentheses.[1] [Reprinted with permission from *Acc. Chem. Res.*, 1968, **2**, 266. Copyright 1968, American Chemical Society.]

is written as $\Psi_0 = \chi_1\chi_2\chi_3$ and the three locally excited state functions are written as $\chi_1'\chi_2\chi_3$, $\chi_1\chi_2'\chi_3$ and $\chi_1\chi_2\chi_3'$, where the numbers 1, 2 and 3 denote each phen molecule, χ is the groundstate function of phen and the primes refer to the excited state functions of phen. Since $\chi_1'\chi_2\chi_3$, $\chi_1\chi_2'\chi_3$ and $\chi_1\chi_2\chi_3'$ are degenerate, the true state is specified as their linear combinations (Table 5.1), which describe the phase relationships between the dipole excitation vectors of each phen molecule. In the $[M(phen)_3]^{n+}$ system, the three coupling modes for the long-axis-polarised transitions are expressed by symmetry notations (Ψ_{A2}, Ψ_{E1}, Ψ_{E2}). The three exciton transitions are $\Psi_0 \rightarrow \Psi_{A2}$, $\Psi_0 \rightarrow \Psi_{E1}$ and $\Psi_0 \rightarrow \Psi_{E2}$ with the last two being degenerate. The coupling modes for the long-axis-polarised $\pi \rightarrow \pi^*$ transitions are shown in Figure 5.2.

The edtm of the $\Psi_0 \rightarrow \Psi_{A2}$ transition, derived from the addition of the linear vectors on the three phen ligands, produces a net linear displacement along the three-fold axis of the molecule normal to the plane of the page. Since the vectors are arranged in the form of a three-bladed propeller, their combined

Table 5.1 Wavefunctions, energies and signs of rotational strength (R) of some Λ configuration *tris*-bidentate octahedral metal complexes. [Reprinted with permission from *Acc. Chem. Res.* 1968, **2**, 266. Copyright 1968, American Chemical Society.]

Compounds	*Wavefunction*	*Energy*	*Sign of R*
	$\Psi_0 = \chi_1\chi_2\chi_3$	0	
$[M(phen)_3]^{n+}$	$\Psi_{A2} = (1/\sqrt{3})\chi_1'\chi_2\chi_3 + \chi_1\chi_2'\chi_3 + \chi_1\chi_2\chi_3')$	$E_A = \Delta E + 2V$	−
$[M(bipy)_3]^{n+}$	$\Psi_{E1} = (1/\sqrt{6})(2\chi_1'\chi_2\chi_3 - \chi_1\chi_2'\chi_3 - \chi_1\chi_2\chi_3')$	$E_E = \Delta E - V$	+
$[M(acac)_3]^{n+}$	$\Psi_{E2} = (1/\sqrt{2})(\chi_1\chi_2'\chi_3 - \chi_1\chi_2\chi_3')$	$E_E = \Delta E - V$	+

ΔE is the transition energy of the long-axis-polarised $\pi \rightarrow \pi^*$ transitions. If the sign in front of V is positive, this transition appears at higher energy, while in the case of negative signals, the transition appears at longer wavelength than ΔE. As a result, in the case of Λ configuration complexes, the CD sign changes from positive to negative in ascending energy. For more details, see ref. 1 and refs cited therein.

effect also produces circulation of charge about the C_3 axis. The direction of the mdtm can be derived using Ampere's rule for solenoids. When the fingers of the right hand are curled in the flow direction, the induced mdtm is aligned in the direction the thumb is pointing. Thus, for a particular absolute configuration (Λ) shown in Figure 5.2, the $\Psi_0 \rightarrow \Psi_{A2}$ transition produces an mdtm directed into the paper, which is the opposite alignment from that of the edtm. A negative CD signal is, therefore, expected for this transition based on the Rosenfeld equation, Table 5.1. In the $\Psi_0 \rightarrow \Psi_{E2}$ transition, two phen molecules are excited. The coupling combination involves a right-handed helical displacement of charge along and around the two-fold axis in the plane of the paper (from left to right). Since the alignment of both the $\boldsymbol{\mu}$ and $\boldsymbol{m}$ are the same, a positively signed CD signal is anticipated. The effect of the $\Psi_0 \rightarrow \Psi_{E1}$ transition is not straightforward, but can be explained by taking the arrows in pairs. The 1,3 and 2,3 pairs produce positive CD intensity, since both $\boldsymbol{\mu}$ and $\boldsymbol{m}$ are aligned in the same direction so that $\boldsymbol{\mu} \cdot \boldsymbol{m} > 0$, while the 1,2 pair involves the anticlockwise helical displacement of charge. In this case, $\boldsymbol{\mu}$ points into the plane formed by the page, while $\boldsymbol{m}$ is aligned in the opposite direction resulting in negative CD intensity since $\boldsymbol{\mu m} < 0$. However, since this is a symmetrical system and the three ligands are identical, the coupling of each pair of ligands leads to the same contribution. A positively signed CD band is observed, since the signal associated with the 1,3 and 2,3 pairs is twice as strong as the negative CD intensity associated with the 1,2 pair. From the "sum rule" that the total observed CD intensity becomes zero if the $\Psi_0 \rightarrow \Psi_{A2}$ transition occurs at the same energy as those of the $\Psi_0 \rightarrow \Psi_{E1}$ and $\Psi_0 \rightarrow \Psi_{E2}$ transitions, the exciton rotational strength should vanish. However, in the Ψ_{A2} coupling mode, the tips or origins of the arrows of the edtms lie in close proximity and have the same sign so they repel each other based on the coulombic interaction, while in the $\Psi_0 \rightarrow \Psi_{E2}$ transition case the tips or origins of the arrows of the edtms have opposite signs and, therefore, attract each other. As a result, the $\Psi \rightarrow \Psi_{A2}$ transition occurs at a higher energy than the degenerate $\Psi_0 \rightarrow \Psi_{E2}$ and $\Psi_0 \rightarrow \Psi_{E1}$ transitions so the CD signal changes from positive to negative sign in ascending energy terms in the spectral region dominated by the long-axis-polarised $\pi \rightarrow \pi^*$ band of phen. Note that the above explanation applies only to the Λ absolute configuration and that, in the Δ configuration, the sign sequence will be the opposite of that described. The short-axis-polarised transitions couple resulting in two transitions, $\Psi_0 \rightarrow \Psi_A$ and the doubly degenerate $\Psi_0 \rightarrow \Psi_E$, Figure 5.2. In the $\Psi_0 \rightarrow \Psi_A$ transition, all arrows point outwards and no net moment is produced. As a result, this transition is forbidden ($\boldsymbol{\mu} = 0$). The two $\Psi_0 \rightarrow \Psi_E$ coupling modes produce a linear displacement of charge so that $\boldsymbol{\mu} \neq 0$, but do not generate any circulation of charge so no exciton-coupled CD intensity is anticipated in this spectral region, since $\boldsymbol{m} = 0$.

Figure 5.3 contains the absorption and CD spectra of $\Lambda - [Ru(phen)_3](ClO_4)_2$. Weak bands in the 19 000–27 000 cm^{-1} region have been assigned to CT transitions from a metal $d\pi$ orbital to a π^* antibonding orbital of the ligand. The 27 000–33 000 cm^{-1} region is dominated by short-axis-polarised ligand transitions, while the intense band centred at 38 000 cm^{-1} is associated with the long-axis-polarised transition. The CD signal in the long-axis-polarised region of the

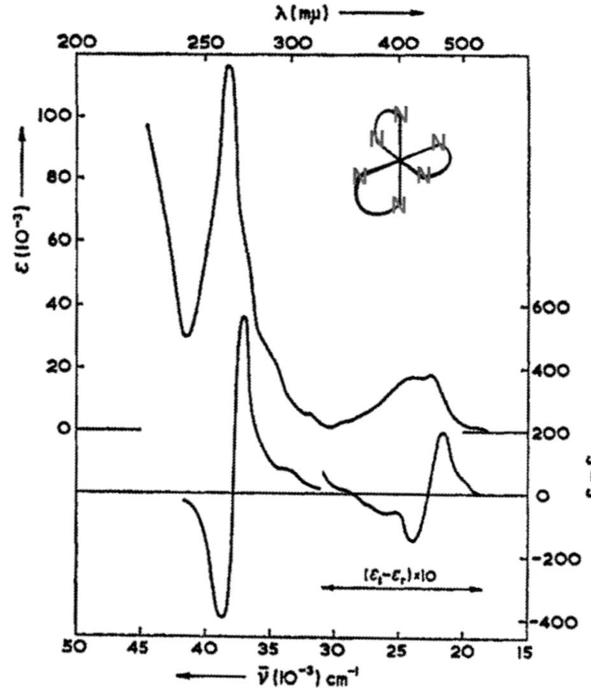

Figure 5.3 Absorption and CD spectra of $\Lambda - [Ru(phen)_3](ClO_4)_2$ in aqueous solution and the predicted absolute configuration.[1] [Reprinted with permission from *Acc. Chem. Res.*, 1968, **2**, 266. Copyright 1968, American Chemical Society.]

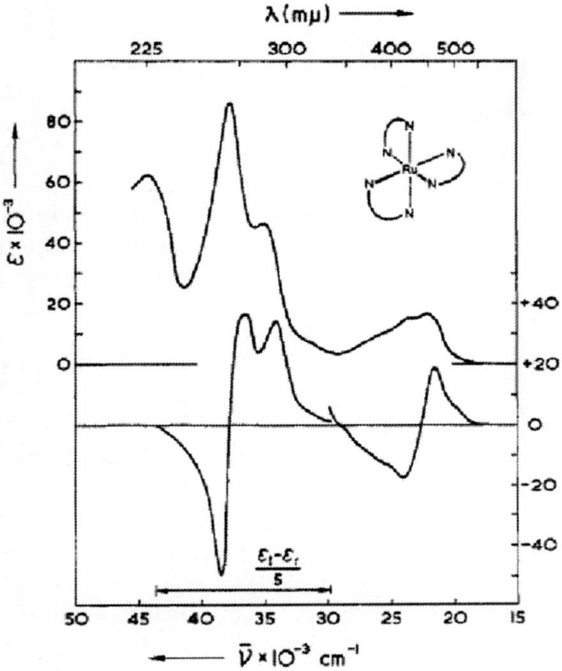

Figure 5.4 Absorption and CD spectra of Λ – [Ru(phen)$_2$bipy](ClO$_4$)$_2$ in aqueous solution and the predicted absolute configuration.[3] [Reprinted with permission from *Inorg. Chem.*, 1968, **7**, 2379. Copyright 1968, American Chemical Society.]

spectrum is more than one order of magnitude stronger than that in the 19 000–33 000 cm^{-1} region and its sign changes from positive to negative in ascending energy, as would be anticipated based on the exciton coupling theory analysis described above.

Figure 5.4 contains the spectra of Λ–[Ru(phen)$_2$bipy](ClO$_4$)$_2$.[3] In this complex, one of the phen ligands of Λ–[Ru(phen)$_3$](ClO$_4$)$_2$ is replaced by a bipy, so that this is a nondegenerate coupled-oscillator system. Although the long-axis-polarised band of bipy lies at 34 500 cm^{-1} rather than 38 000 cm^{-1}, the CD spectral pattern in this region is still consistent with exciton coupling theory. It is therefore believed that the absolute configuration of *tris*-bidentate compounds can be determined from the CD sign in the long-axis-polarised transition region. *If the CD signal appears to be strongly positive at lower energy and strongly negative at higher energy, the molecule has a Λ–absolute configuration.* Λ–[Ru(phen)$_2$(py)$_2$](ClO$_4$)$_2$ and Λ–[Ru(phen)(bipy)$_2$](ClO$_4$)$_2$ complexes exhibit plus-to-minus CD patterns in ascending energy terms in the long-axis-polarised band region, providing further evidence for the relationship between the observed CD sign sequence and the absolute configuration.

5.2 CD spectra in the $d \rightarrow d$ Transition Region

The $d \rightarrow d$ bands generally occur at longer wavelengths than charge transfer and intraligand bands. During the course of a $d \rightarrow d$ transition, the electron in the initial d orbital is rotated over into another d orbital. Since no linear motion of charge is generated, $d \rightarrow d$ transitions are electric dipole forbidden, so their absorption coefficients are generally several orders of magnitude weaker than those of electric dipole allowed $\pi \rightarrow \pi^*$ transitions. The CD bands associated with $d \rightarrow d$ transitions are known to depend both on the symmetry of the metal complex and the directly ligated atoms, and on the alignment of the intrinsic mdtms associated with the magnetic dipole allowed $d \rightarrow d$ transitions.[4]

5.2.1 $[Co(en)_3]^{3+}$ and $[Cr(en)_3]^{3+}$

The electronic absorption spectrum of $[Co(en)_3]^{3+}$, Figure 5.5,[4] contains bands at *ca.* 470 nm (band I) and 340 nm (band II) with absorption coefficients of less than 100 and an intense CT band at *ca.* 210 nm with an ε value of *ca.* 150 000, Figure 5.6.[5] Figure 5.7 provides the assignments and nomenclature of the $d \rightarrow d$ transitions in the context of O_h and D_3 symmetry complexes. In octahedral complexes, T_{1g} symmetry transitions are magnetic dipole allowed and electric dipole forbidden, while T_{2g} symmetry transitions are both magnetic and electric dipole forbidden. The $[Co(en)_3]^{3+}$ ion has D_3 symmetry, Figure 5.5. Band I arises from the T_{1g} transitions, Figure 5.7, while band II arises from the T_{2g} transitions. As would be anticipated on this basis, the CD intensity associated with band II is much weaker than that of band I. In addition, the CD spectrum recorded in solution demonstrates that band I consists of two components which appear to be associated with the $A_1 \rightarrow A_2$ and $A_1 \rightarrow E$ transitions. These

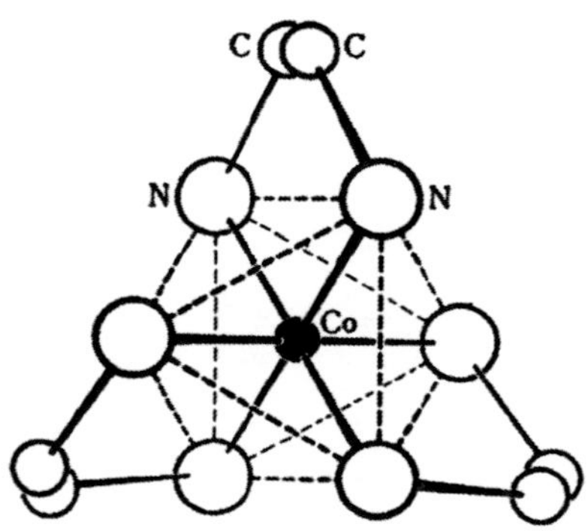

Figure 5.5 The structure of the D(+)-$[Co(en)_3]^{3+}$ complex ion. Six nitrogen atoms form a slightly distorted octahedron around the central cobalt atom. The five-membered cobalt-ethylenediamine ring is not planar; each ethylenediamine ligand adopts the so-called 'gauche' form.[4] [Reprinted with permission from *Bull. Chem. Soc. Jpn.*, 1957, **30**, 795. Copyright 1957, The Chemical Society of Japan.]

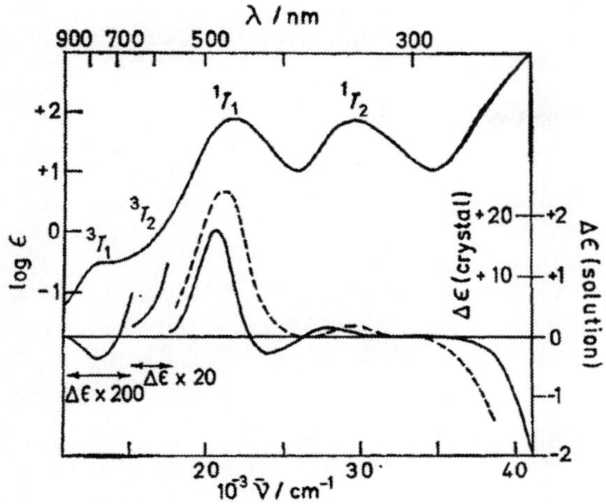

Figure 5.6 Absorption (broken line) and CD spectra (broken dotted line) in water and CD spectra of crystalline 2{Λ – [Co(en)$_3$]Cl$_3$}NaCl.6H$_2$O (solid line). The CD spectrum of hexagonal crystals of 2{Λ – [Co(en)$_3$]Cl$_3$}NaCl.6H$_2$O was measured while adjusting the direction of light propagation parallel to the trigonal axis of [Co(en)$_3$]$^{3+}$ ion.[5b] [Reprinted with permission from *J. Chem. Soc., Dalton Trans.* 1977, 937. Copyright 1977, Royal Society of Chemistry.]

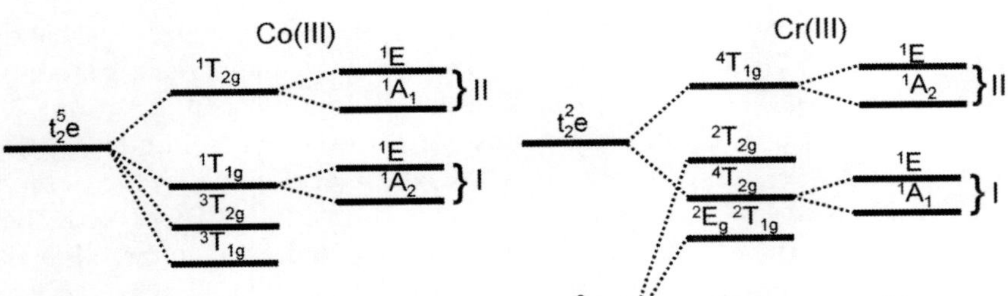

Figure 5.7 Assignment of the $d \rightarrow d$ transitions of Co(III) and Cr(III) under O_h and D_3 symmetry to bands labelled I and II associated with the $^1T_{1g}$ and $^1T_{2g}$ excited states, Figure 5.6.[5]

transitions are both magnetic dipole allowed, Table 5.2. The latter has two mutually orthogonal components polarised in the plane perpendicular to the trigonal axis of the [Co(en)$_3$]$^{3+}$ ion, while the former is polarised parallel to the axis. In order to determine the band assignments, the CD spectrum of hexagonal crystals of 2{Λ – [Co(en)$_3$]Cl$_3$}NaCl.6H$_2$O was measured while

Table 5.2 Selection rules in D_3 symmetry in which the C_3 symmetry axis lies on the z-axis.

	A_1	A_2	E
A_1	—	z	x/y
A_2	z	—	x/y
E	x/y	x/y	$z,x/y$

maintaining the direction of light propagation was kept parallel to the trigonal axis. Under these circumstances, only the E symmetry component should be detected. As depicted by the solid line in Figure 5.6, only positive intensity was detected in this region, so the positively signed CD band in the solution spectrum can be assigned to the $A_1 \rightarrow E$ transition, while the weak negatively signed trough of intensity can be assigned to the $A_1 \rightarrow A_2$ transition. Since the CD band associated with the $A_1 \rightarrow E$ transition was found to be more intense than that associated with the $A_1 \rightarrow A_2$ transition for a significant number of complexes, this assignment is often referred to as the 'dominant E-band rule'. Although this can be explained on a theoretical basis,[6] care must be taken in this regard, since the solution spectrum is thought to be produced by the summation of the strong positive CD signal for the E symmetry component and the negative CD signal for an A_2 symmetry component of similar magnitude but opposite sign, which can be detected separately when crystals are used. The CD signal observed in solution, therefore, arises from the cancellation of two intense bands with only a small percentage of the total signal being observed experimentally.

In the case of (+)-$[Cr(en)_3]^{3+}$, a very sharp, positively signed CD band is observed for band I but no negative component can be identified, Figure 5.8. This complex has, therefore, been considered to adopt a Λ configuration.[7] The band at 21 900 cm^{-1} (positive CD) is assigned to the first spin allowed transition to the ^4E ($^4T_2$) state. Three narrow bands of weak intensity are observed in the CD spectrum, due to spin-forbidden transitions. These have been assigned, taking the splitting associated with the spin-orbit interaction into account, to the following transitions; 14 900 cm^{-1} (positive CD signal) to E, 2A (^2E), at 15 200 cm^{-1} (negative CD signal) to E_b ($^2T_1$), and at 15 500 cm^{-1} (positive CD signal) to 2A ($^2T_1$).

5.2.2 Near-IR Absorption and CD Spectra of Ferrocyctochrome c: *d* → *d* Transitions

In the late 1960s, the $d \rightarrow d$ bands of transition metal porphyrin complexes were identified for the first time on the basis of an analysis of the electronic absorption and CD spectra of ferrocytochrome *c*.[8] This system still provides a good example of how to identify these $d \rightarrow d$ bands in experimental spectra. Figure 5.9 contains a schematic structure of the haem group in this enzyme, which contains of a low spin d^6 diamagnetic Fe(II) complex. The iron is

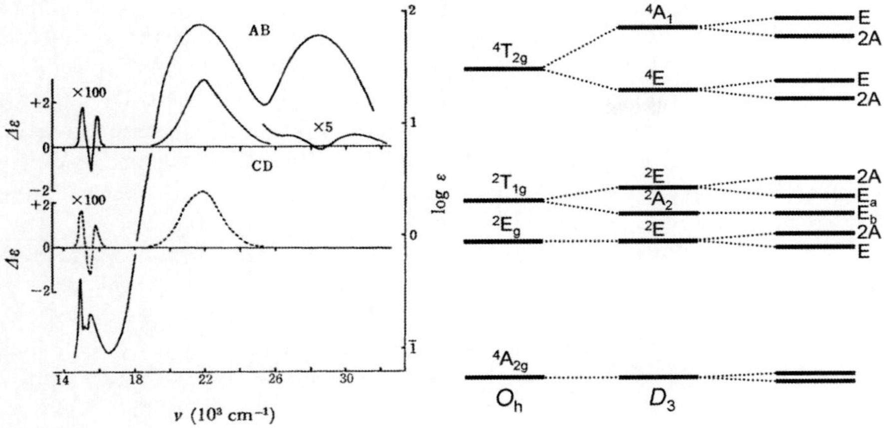

Figure 5.8 Absorption and CD spectra of (+)-$[Cr(en)_3]^{3+}$ in water (left). Solid and dashed lines denote CD spectra recorded in water and 0.4 M Na_2SeO_3 aqueous solution, respectively. Energy levels of the Cr(III) ion in the absence (O_h, D_3) and presence of a spin–orbit interaction (right).[7] [*Bull. Chem. Soc. Jpn.*, 1970, **43**, 1100. Copyright 1970, The Chemical Society of Japan.]

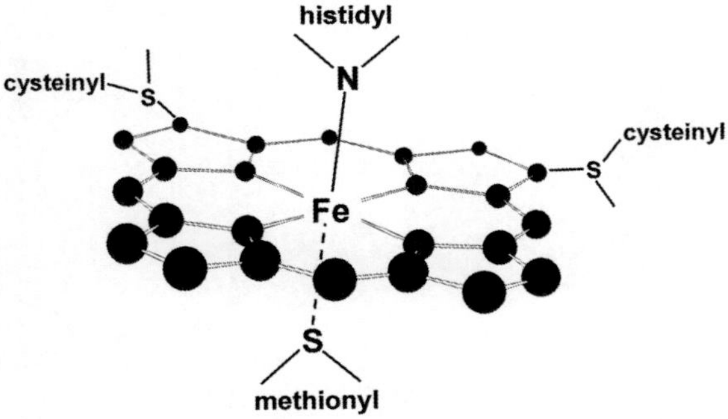

Figure 5.9 Schematic structure of the haem group in ferrocytochrome *c*.[8]

hexacoordinated in a roughly tetragonal manner to the four pyrrole nitrogens of the porphyrin ring, with the two axial ligands provided by an imidazole nitrogen of a histidyl residue and a dialkyl sulfur of a methionyl residue. In diamagnetic $3d^6$ octahedral complexes ($^1A_{1g}$ groundstate) the two lowest crystal field singlet states belong to the T_{1g} and T_{2g} irreducible representations of the O_h point group. The $^1T_{1g}$ state lies at lower energy than the $^1T_{2g}$ state. Transitions to the higher energy $^1T_{2g}$ state are magnetic dipole forbidden, while transitions to the lower energy $^1T_{1g}$ states are magnetic dipole allowed. Both of these transitions are electric dipole forbidden in strict O_h symmetry. Thus, if a CD signal is to be detected upon a lowering of the molecular symmetry, transitions to the new crystal field states derived from the $^1T_{1g}$ state should be distinguishable from other bands by their relatively large anisotropy factors (four times the rotational strength divided by the dipole strength). This value is larger for bands with strong CD intensity, since a contribution of the magnetic dipole moment is required because the rotatory strength is based on $\boldsymbol{\mu} \cdot \boldsymbol{m}$, equation (1.17). The first step in locating the $d \rightarrow d$ band is, therefore, to derive the anisotropy factor value from the experimental spectra using spectral band deconvolution.

Figure 5.10 contains the electronic absorption and CD spectra of horse heart ferrocytochrome *c*. The very intense tail of absorbance beyond 17 500 cm^{-1} is associated with the porphyrin $\pi \rightarrow \pi^*$ transitions. In this region, the anisotropy factor is low, *ca.* 0.0002, as would be expected for magnetic dipole forbidden $\pi \rightarrow \pi^*$ transitions. However, in the region between 10 000 and 17 500 cm^{-1}, the anisotropy factor is about one order of magnitude larger. This identifies some of these near-IR bands as being magnetic dipole allowed. The low dipole strengths of the absorption bands in this region are also consistent with this conclusion. Three and four components are identified in the absorption and CD spectra, respectively, by band deconvolution analysis. Two CD bands were required corresponding to the absorption component at the longest wavelength (band I). The anisotropy factor of band I was 0.065, which is more than 30 times larger than that of the $\pi \rightarrow \pi^*$ transitions. This value by itself is large

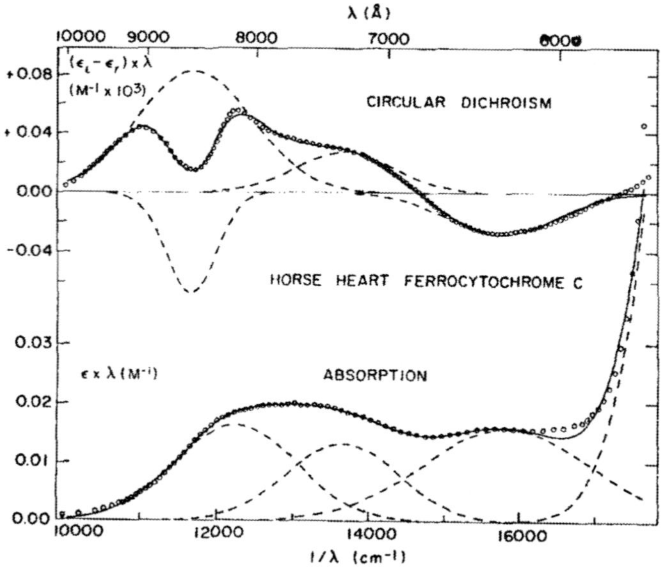

Figure 5.10 Absorption and CD spectra of horse heart ferrocytochrome *c* with resolution into Gaussian-shaped bands.[8] [Reprinted with permission from *J. Chem. Phys.*, 1969, **51**, 4502. Copyright 1969, American Institute of Physics.]

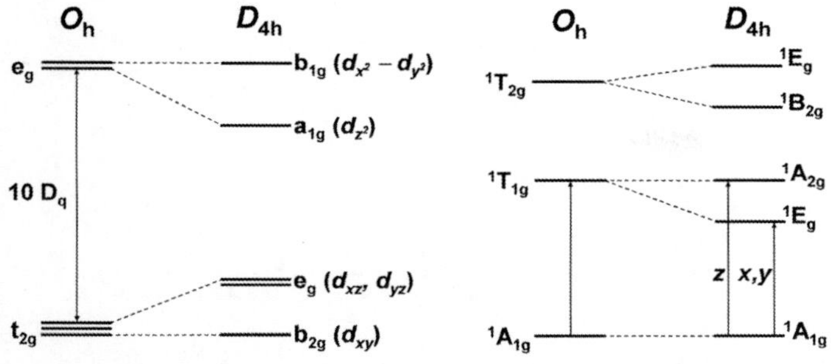

Figure 5.11 3*d* orbitals in an octahedral and tetragonal crystal field (left) and singlet states of a low spin d^6 ion in an octahedral and tetragonal crystal field (right).[8]

enough to conclude that this is a $d \rightarrow d$ band, but the detailed rationale is as follows. The splitting of the 3*d* orbitals of a low spin d^6 ion in an O_h or D_{4h} crystal field results in the magnetic dipole allowed transitions to singlet excited states described in Figure 5.11. As explained previously, in Section 1.2., for a transition to be optically active it must possess edtm and mdtm vectors that are parallel or antiparallel. Since the $^1A_g \rightarrow {}^1E_g$ ligand field transition is expected to be polarised in the porphyrin plane, only in-plane polarised electric dipole intensity can contribute to the rotational strength of this $d \rightarrow d$ transition. The $^1A_g \rightarrow {}^1A_{2g}$ $d \rightarrow d$ transition is magnetic dipole allowed, but is predicted to be polarised normal to the porphyrin plane, Figure 5.11. As a result, only the $^1A_g \rightarrow {}^1E_g$ $d \rightarrow d$ transition can borrow intensity from the nearby strong in-plane (electric) polarised porphyrin $^1A_g \rightarrow {}^1E_g$ $\pi \rightarrow \pi^*$ transitions, giving rise to a larger anisotropy factor. Thus, the lowest energy band is assigned to a $^1A_g \rightarrow {}^1E_g$ $d \rightarrow d$ transition, which is in close agreement with the prediction made on the basis of MO calculations that the lowest-lying crystal field excited singlet state has ^1E_g symmetry.

References

1. B. Bosnich, *Acc. Chem. Res.*, 1968, **2**, 266.
2. J. Hidaka and B. E. Douglas, *Inorg. Chem.*, 1964, **3**, 1180.
3. B. Bosnich, *Inorg. Chem.*, 1968, **7**, 2379.
4. Y. Saito, K. Nakatsu, M. Shiro and H. Kuroya, *Bull. Chem. Soc. Jpn.*, 1957, **30**, 795.
5. (a) A. J. McCaffery and S. F. Mason, *Mol. Phys.*, 1963, **6**, 359; (b) S. F. Mason and B. J. Peart, *J. Chem. Soc., Dalton Trans.*, 1977, 937.
6. P. E. Shipper and A. Rodger, *Chem. Phys.*, 1986, **109**, 173.
7. (a) S. Kaizaki, J. Hidaka and Y. Shimura, *Bull. Chem. Soc. Jpn.*, 1970, **43**, 1100. (b) S. Kaizaki, J. Hidaka and Y. Shimura, *Inorg. Chem.*, 1973, **12**, 142.
8. W. A. Eaton and E. Charney, *J. Chem. Phys.*, 1969, **51**, 4502.

CHAPTER 6

Circular Dichroism Induced by Optically Active Binaphthyl

1,1'-Binaphthyl dimers, oligomers and polymers have been used for molecular recognition, asymmetric catalysis and as new materials in a variety of different systems.[1,2] From the standpoint of CD spectroscopy, only the exciton coupling between pairs of naphthalene chromophores has been studied in depth.[3–5] 1,1'-Binaphthyl dimers can also be used to carry out analyses based on induced CD (ICD) spectroscopy, if they are introduced properly to another chromophore.[6] Ideally this chromophore should be a large, symmetrical and intrinsically achiral molecule, preferably with strong absorption in regions of the spectrum where binaphthyl does not absorb significantly. In addition, for ease of analysis, it is important to link the binaphthyl moiety at the lowest possible distance from this chromophore in a fixed orientation, in order to increase the mutual interaction and hence the intensity of the ICD signal. From this standpoint, D_{4h} symmetry phthalocyanines are probably the best candidates for demonstrating the potential utility of this approach, since they are rigid, square planar molecules, with very strong absorption bands, usually referred to as the Q and B1 (or Soret) bands, in the 600–700 and 300–400 nm regions, respectively, two spectral regions where binaphthyl does not absorb significantly.

6.1 Chiral Binaphthyl-Induced CD in Phthalocyanines

Porphyrins and phthalocyanines have characteristic $\pi \rightarrow \pi^*$ bands in the UV-visible region of the electronic absorption spectrum that are in-plane polarised and achiral. As was explained in Section 1.2.2, the Kuhn-Kirkwood coupled oscillator and "CD stealing" mechanisms, equations (1.22–1.24), both have to be taken into consideration in any theoretical description of the ICD of achiral

Circular Dichroism and Magnetic Circular Dichroism Spectroscopy for Organic Chemists
By Nagao Kobayashi, Atsuya Muranaka and John Mack

Published by the Royal Society of Chemistry, www.rsc.org

electric dipole transitions.[7–9] If an achiral chromophore is introduced, the rotatory strength can only arise when the perturber has an edtm which is skewed with respect to the moment on the ring system.[10,11] When the perturber exhibits strong CD, as is the case with 1,1'-binaphthyl dimers, another induction channel is opened, which can be transferred directly to the ring system through a "CD stealing" mechanism. In the following theoretical treatment the phthalocyanine (Pc) and chiral binaphthyl substituent are designated by *i* and *j*, respectively. A third possible mechanism would involve coupling between the mdtm of the central chromophore and the edtms of the peripheral substituents. In metallophthalocyanines with D_{4h} symmetry, transitions to both the B1 and Q excited states transform as E_u, however, while magnetic moments require $E_g + A_g$ symmetries. Hence, neither of the Pc transitions carries an intrinsic magnetic moment, so this third mechanism is symmetry-forbidden in the context of a coupled chromophore model and can be safely neglected.

Figure 6.1 depicts two types of binaphthyl-substituted Pc structure. Two or four binaphthyls are linked to Pcs *via* sulfur or oxygen atoms and the mutual

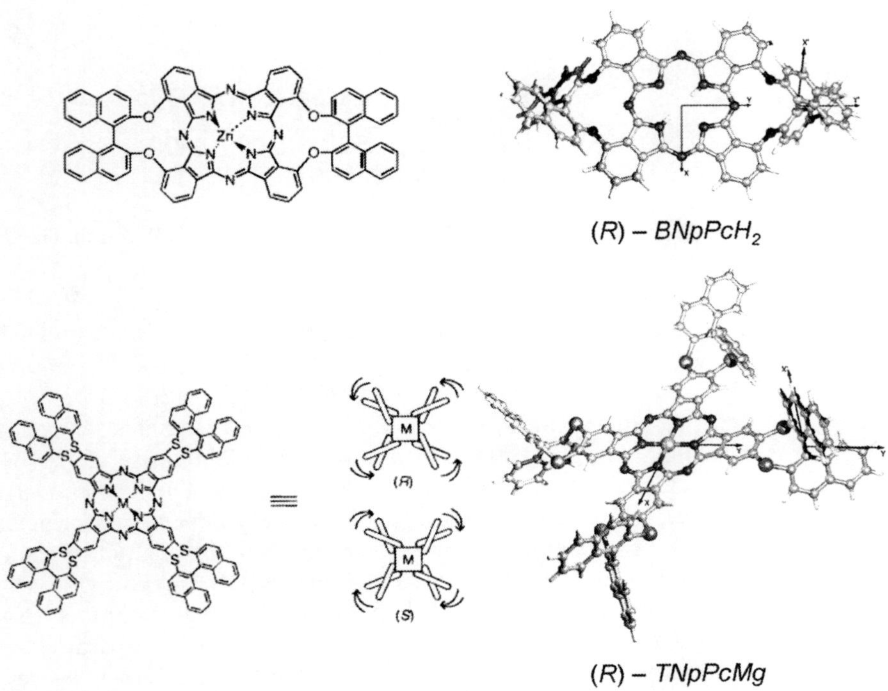

Figure 6.1 Structures (left) and optimised geometries (right) of free base and magnesium phthalocyanine structures with two ($BNpPcH_2$) or four (TNpPcMg) binaphthyl-substituents.[6] In the former case the binaphthyl moieties are linked to neighbouring benzo groups of the phthalocyanine ring by oxygen atoms, while in the latter case a binaphthyl moiety is attached to each benzo group by a pair of sulfur atoms.

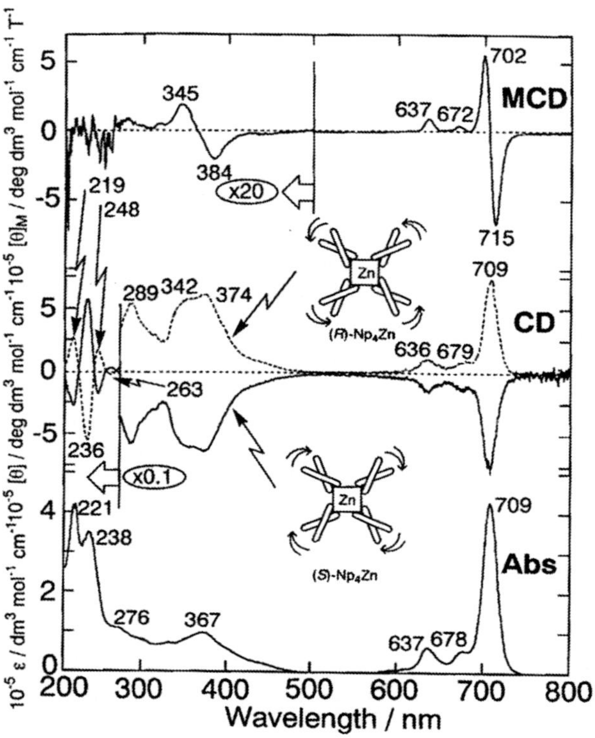

Figure 6.2 The electronic absorption, CD and MCD spectra of TNpPcZn in THF.[6]

alignments of the Pc and binaphthyl rings are firmly fixed. Figures 6.2 and 6.3 contain the optical spectra of the Zn(II) complexes. An intense Q band is observed at 709 nm, Figure 6.2, which is associated with a transition to a doubly degenerate excited state. The morphology of the CD signal is similar to that of the corresponding absorption band, although the intensity of the B1 band is stronger relative to the Q band in the CD spectrum. The (*R*)- and (*S*)-binaphthyls can be confirmed to be left-handed and right-handed conformers, respectively, based on the exciton coupling of the signals of the binaphthyl units in the 200–245 nm region. The sign of the ICD band of (*R*)-binaphthyl-linked Pc in the 270–750 nm region was found to be positive, while that of (*S*)-binaphthyl-linked Pc was found to be negative. The experimental parameters for the Q and B1 bands were obtained by fitting the data using Gaussian curves, Tables 6.1 and 6.2. Figure 6.4 summarises the geometrical parameters required for the analysis, which were obtained from the optimised structures.

The first step in the construction of a theoretical model for this system is to describe the optical activity of the isolated binaphthyl group. A simple Kuhn-Kirkwood model is used, since it determines how the $\boldsymbol{\mu}_j$ and $\boldsymbol{m}_j$ moments are orientated and provides an indication of the level of understanding that can be

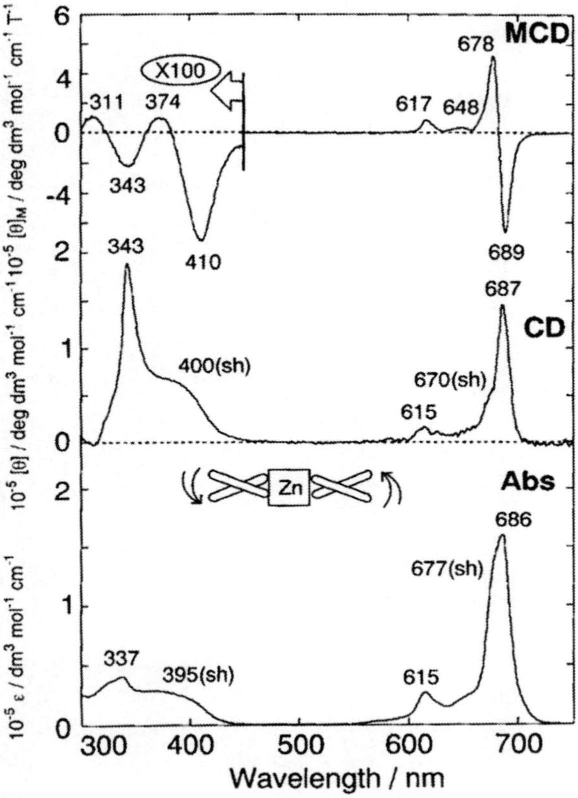

Figure 6.3 The electronic absorption, CD and MCD spectra of BNpPcZn in DMF.[6]

Table 6.1 Results of Gaussian fit analysis (absorption spectra).[6]

complex		ν_{max}(cm^{-1})	ε_{max}(dm^3/mol cm)	$\Delta\nu_{1/2}$(cm^{-1})	*f*	$\lvert\boldsymbol{\mu}\rvert$(D)
BNpPcZn	Q(0,0)	14 611	158 313	445.5	0.3495	7.128
		14 866	25 845	212.0		
TNpPcZn	Q(0,0)	14 162	421 715	409.4	0.7939	10.919
	UV	43 070	339 359	4765.6	7.4356	19.162
		47 243	369 549	3414.5	5.8015	16.161

Table 6.2 Results of Gaussian fit analysis (CD spectra).[6]

complex		ν_{max}(cm^{-1})	θ_{max}(*deg* dm^3/mol)	$\Delta\nu_{1/2}$(cm^{-1})	R(Dμ_B)
(*R*)–BNpPcZn	Q(0,0)	14 544	9.63	264.0	0.2740
		14 856	2.84	311.7	
(*R*)–TNpPcZn	Q(0,0)	14 191	51.56	417.2	1.7188
	UV	43 696	−122.0	2315.6	22.5748
		47 373	50.70	2983.7	12.0895

	$BNpPcH_2$	TNpPcMg
R_{12}	8.62 Å	11.14 Å
r_{12}	4.33 Å	4.30 Å
α	82.58°	73.74°
θ	55.42°	67.72°

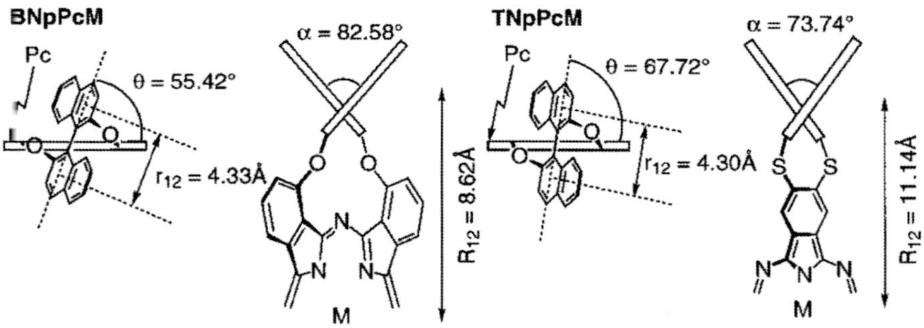

Figure 6.4 Key geometrical parameters and optimised values for the di- and tetra-binaphthyl-substituted Pcs.[6]

anticipated. The optical activity of binaphthyl is associated with the long-axis-polarised B_b transition of the naphthyl units. Naphthalene also has a short-axis-polarised L_a band below 40 000 cm^{-1} (250 nm), which is one order of magnitude less intense than the B_b band. This transition has no intrinsic chirality and therefore cannot contribute to the "CD stealing" mechanism. The L_b band is three orders of magnitude less intense than the B_b band and can therefore also be safely neglected. Although, in principle, there could be a minor contribution of ICD intensity from the Kuhn-Kirkwood coupled dipole model, the actual geometry, which is based on an angle, θ, between its direction of polarisation and the Pc plane of close to 90°, Figure 6.4, does not favour this intensity mechanism. If the *R* configuration is considered, $\boldsymbol{\mu}_1$ and $\boldsymbol{\mu}_2$ denote the individual long-axis-polarised moments on the naphthyl rings. These moments form a twist arrangement with a "propeller angle" which is equal to the dihedral angle, α, between the two naphthyl planes. Figure 6.5 provides a view of the moments, $\boldsymbol{\mu}_1$ and $\boldsymbol{\mu}_2$, in the local Cartesian axis frame (x', y', z') with the origin placed in the centre of the binaphthyl bridge. In the context of this axis frame, the moments can be described as follows:

$$\boldsymbol{\mu}_1 = \boldsymbol{\mu}_N\{\cos(\alpha/2)\mathbf{e}_{y'} - \sin(\alpha/2)\mathbf{e}_{z'}\} \quad (6.1)$$

$$\boldsymbol{\mu}_2 = \boldsymbol{\mu}_N\{\cos(\alpha/2)\mathbf{e}_{y'} + \sin(\alpha/2)\mathbf{e}_{z'}\} \quad (6.2)$$

where μ_N denotes their length and the **e** symbols are unit vectors along the specified directions, $\mathbf{r}_1$ and $\mathbf{r}_2$, directed to the centres of the bridged benzene

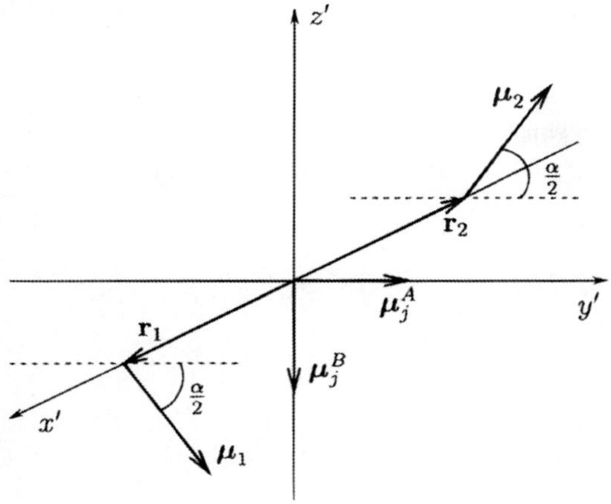

Figure 6.5 View of the individual long-axis-polarised moments on the naphthyls in the *R*-configuration in the local (x', y', z') Cartesian axis frame with the origin at the centre of the binaphthyl bridge.[6]

rings. The distance $|\mathbf{r}_1 - \mathbf{r}_2|$ is denoted as r_{12}. The exciton eigenstates correspond to the normalised sum and difference combinations of the two transitions, which are respectively symmetrical (A) and antisymmetrical (B) with respect to the two-fold axis through the central bridge. The corresponding transition moments are, therefore, given by:

$$\boldsymbol{\mu}_j^{A} = (\boldsymbol{\mu}_1 + \boldsymbol{\mu}_2)/\sqrt{2} \tag{6.3}$$

$$\boldsymbol{\mu}_j^{B} = (\boldsymbol{\mu}_1 - \boldsymbol{\mu}_2)/\sqrt{2} \tag{6.4}$$

and the separation between the two states amounts to:

$$E_j^{A} - E_j^{B} = 2V_{12} \tag{6.5}$$

where V_{12} represents the interaction element which transfers the excitation. For 1,1'-binaphthyls, the symmetrical A component has been demonstrated to be the higher energy state[12] when the dihedral angle α is close to 90°. On the basis of equation (6.3), the edtm for this state can be expressed as:

$$\boldsymbol{\mu}_j^{A} = \sqrt{2}\boldsymbol{\mu}_N \cos(\alpha/2)\mathbf{e}_{y'} \tag{6.6}$$

The coupling of the two naphthyl oscillators in this state also generates a mdtm, $\boldsymbol{m}_j^{A}$, which is given by:

$$\boldsymbol{m}_j^{A} = (\pi i \nu_N/\sqrt{2})(\mathbf{r}_1 \times \boldsymbol{\mu}_1 + \mathbf{r}_2 \times \boldsymbol{\mu}_2) = (\pi i \nu_N \boldsymbol{\mu}_N/\sqrt{2}) r_{12} \sin(\alpha/2)\mathbf{e}_{y'} \tag{6.7}$$

The mdtm and edtm of the A state are, therefore, both directed along the C_{2y} axis, resulting in a non-zero rotatory strength for the transition to the higher energy state:

$$R_j^{A} = \mathrm{Im}(\boldsymbol{m}_j^{A} \cdot \boldsymbol{\mu}_j^{A}) = (\pi\nu_N\boldsymbol{\mu}_N^{2}/2)\mathrm{r}_{12}\sin\alpha \tag{6.8}$$

In a similar way, for the lower energy B state:

$$\boldsymbol{\mu}_j^{B} = -\sqrt{2}\boldsymbol{\mu}_N \sin(\alpha/2)\mathbf{e}_{z'} \tag{6.9}$$

$$\boldsymbol{m}_j^{B} = (\pi i\nu_N/\sqrt{2})(\mathbf{r}_1 \times \boldsymbol{\mu}_1 - \mathbf{r}_2 \times \boldsymbol{\mu}_2) = (\pi i\nu_N\boldsymbol{\mu}_N/\sqrt{2})\mathrm{r}_{12}\cos(\alpha/2)\mathbf{e}_{z'} \tag{6.10}$$

Both moments are thus directed along the z' direction. The rotatory strength for the lower state is:

$$R_j^{B} = \mathrm{Im}(\boldsymbol{m}_j^{B} \cdot \boldsymbol{\mu}_j^{B}) = -(\pi\nu_N\boldsymbol{\mu}_N^{2}/2)\mathrm{r}_{12}\sin\alpha \tag{6.11}$$

On this basis, a minus-to-plus CD pattern is anticipated in the long-axis-polarised B_b transition region (245–200 nm region) for the *R* configuration of binaphthyl, as is indeed observed, Figure 6.2. In addition, it should be noted that the "sum rule" is obeyed, since equations (6.8) and (6.11) make it clear that, $R_j^{A} + R_j^{B} = 0$.

The second step is to consider the CD intensity that is induced in the Q and B1 bands of Pc by the binaphthyl substituents. It is assumed that the metal Pc complex is an approximately square-planar chromophore, with one binaphthyl substituent placed on the positive *y*-axis. Since both the B1 and Q bands arise from doubly degenerate transitions, they can split into two components which are symmetrical and antisymmetrical with respect to a two-fold symmetry axis which coincides with the *y*-axis. These components can be represented as vectors that are orientated along the positive *y*- and *x*-axes, as indicated in Figure 6.6. An overview of all of the possible coupling is presented in Figure 6.7. Because of the long intercentre distance, R_{12}, the interaction between the chromophore and substituent does not result in a significant lifting of the degeneracy of the Q and B1 bands. The observed ICD of these transitions thus corresponds to the superposition of the rotatory strengths associated with the two components. If these have opposite signs, a pronounced net CD signal is anticipated only if there is a clear difference in magnitude. Such differences arise on symmetry grounds. Selection rules prevent interaction between components with different symmetries with respect to the two-fold axis. For this reason, the interaction mechanism for the A and B symmetry components must be examined separately.

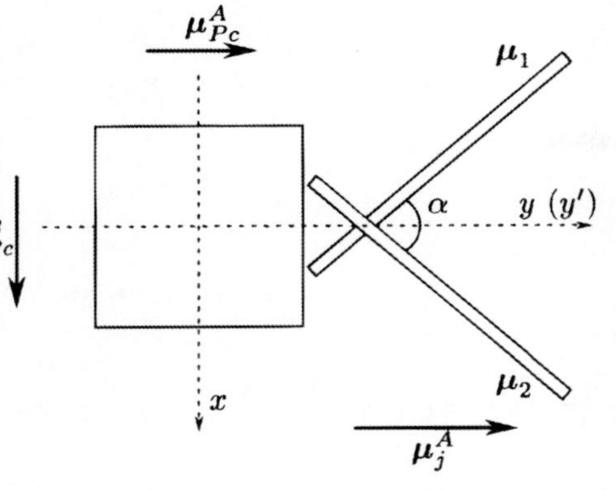

Figure 6.6 The edtms of the phthalocyanine chromophore and the A and B symmetry components of the binaphthyl moiety.[6] The B1 and Q bands arise from doubly degenerate transitions, which can split into two components which are symmetrical and antisymmetrical with respect to the two-fold symmetry axis, which coincides with the y-axis.

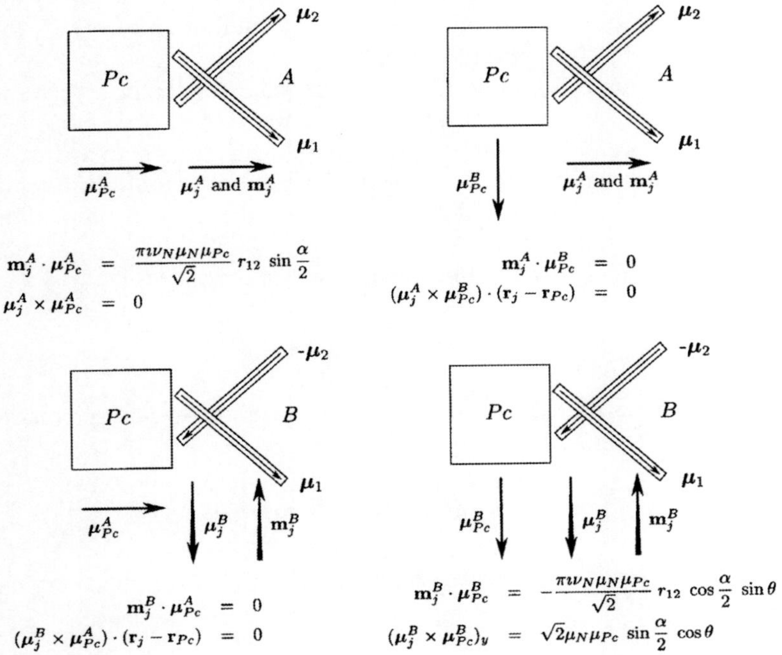

Figure 6.7 Overview of all possible couplings of the edtms of the phthalocyanine chromophore and the edtms and mdtms of the binaphthyl moiety.[6]

Given that the edtm and mdtm of A symmetry transitions are polarised along the *y*-axis and are, therefore, aligned, their vector product will be zero and induction *via* the coupled oscillator mechanism, therefore, remains inactive.

$$R_1^{A} = 0 \tag{6.12}$$

In contrast, the "CD stealing" channel is allowed, since the magnetic moment of an A symmetry transition on the substituent is also aligned along the two-fold direction and can, therefore, couple with the edtm on the Pc ligand. The dipole-dipole interaction term for the A transition is given by:

$$V_{ij}^{A} = -(\sqrt{2}\boldsymbol{\mu}_{Pc}\boldsymbol{\mu}_{N}/2\pi\epsilon_0 R_{12}^{3})\cos(\alpha/2) \tag{6.13}$$

where μ_{Pc} represents the length of one in-plane component of the transition moment of the Pc chromophore. The induced rotatory strength of the A symmetry component on the Pc ligand then becomes:

$$R_2^{A} = \{\nu_{Pc}\nu_{N}\boldsymbol{\mu}_{Pc}^{2}\boldsymbol{\mu}_{N}^{2}r_{12}/2\epsilon_0 h(\nu_N^{2} - \nu_{Pc}^{2})R_{12}^{3}\}\ \sin\alpha \tag{6.14}$$

The transition frequency of the substituent was approximated as ν_N, thus neglecting the exciton splitting of the binaphthyl bands. By inserting the appropriate values for the transition moment ($\boldsymbol{\mu}_{Pc}$) and frequency (ν_{Pc}), R_2^{A} can be obtained for both the B1 and Q bands.

When the B symmetry component is considered, the geometric relationships between the central and peripheral transitions are different from those of the A component. The central B component edtm, $\boldsymbol{\mu}_{PcB}$, lies in the plane of the Pc along $\mathbf{e}_x$, while the edtm of the B symmetry transition on the substituents, $\boldsymbol{\mu}_{jB}$, is also perpendicular to the two-fold axis, but along $-\mathbf{e}_{z'}$. A view along the *y*-direction is shown in Figure 6.8.

The scalar product between these directions can be expressed as:

$$\mathbf{e}_x\cdot(-\mathbf{e}_{z'}) = \cos(\pi/2 - \theta) = \sin\theta \tag{6.15}$$

where the angle θ is that shown in Figure 6.4. The dipole-dipole interaction for the B transition then becomes:

$$V_{ij}^{B} = (\sqrt{2}\boldsymbol{\mu}_{Pc}\boldsymbol{\mu}_{N}/4\pi\epsilon_0 R_{12}^{3})\ \sin(\alpha/2)\ \sin\theta \tag{6.16}$$

The vector term in the coupled oscillator mechanism is then obtained as follows:

$$\begin{aligned}(\mathbf{r}_j - \mathbf{r}_i)\cdot(\boldsymbol{\mu}_j \times \boldsymbol{\mu}_i) &= -\sqrt{2}R_{12}\boldsymbol{\mu}_{Pc}\boldsymbol{\mu}_N \sin(\alpha/2)\ \mathbf{e}_y\cdot[\mathbf{e}_{z'} \times \mathbf{e}_x] \\ &= \sqrt{2}R_{12}\boldsymbol{\mu}_{Pc}\boldsymbol{\mu}_N \sin(\alpha/2)\cos\theta \end{aligned} \tag{6.17}$$

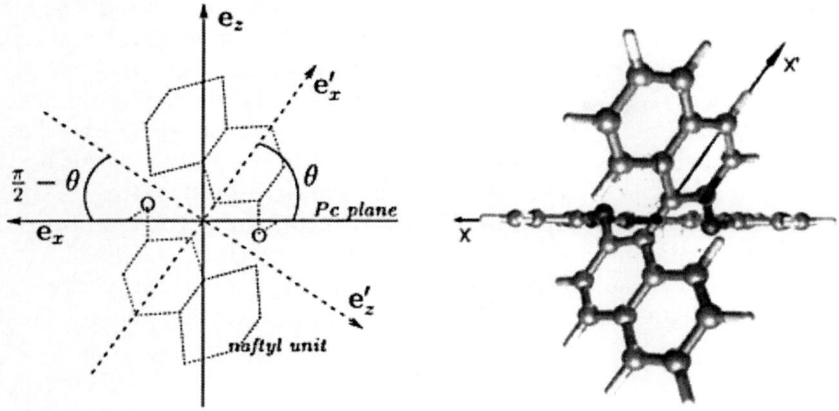

Figure 6.8 View along the *y*-axis. The edtm of the B symmetry transition on the substituents, μ_{jB}, is aligned perpendicular to the two-fold symmetry axis along the $-\mathbf{e}_{z'}$ axis.[6]

Similarly, the scalar product in the CD intensity stealing mechanism is given by:

$$\begin{aligned}\boldsymbol{\mu}_i{}^B \cdot \boldsymbol{m}_j{}^B &= (\pi i \nu_N \boldsymbol{\mu}_{Pc} \boldsymbol{\mu}_N/\sqrt{2}) r_{12} \cos(\alpha/2)\ \mathbf{e}_x \cdot \mathbf{e}_{z'} \\ &= -(\pi i \nu_N \boldsymbol{\mu}_{Pc} \boldsymbol{\mu}_N/\sqrt{2}) r_{12} \cos(\alpha/2) \sin\theta \qquad (6.18)\end{aligned}$$

The two mechanisms contributing to the rotatory strength of the B type transitions can thus be described as:

$$R_1{}^B = -\{\nu_{Pc}\nu_N \boldsymbol{\mu}_{Pc}{}^2 \boldsymbol{\mu}_N{}^2/2\epsilon_0 h(\nu_N{}^2 - \nu_{Pc}{}^2) R_{12}{}^2\} \sin^2(\alpha/2) \sin 2\theta \qquad (6.19)$$

$$R_2{}^B = \{\nu_{Pc}\nu_N \boldsymbol{\mu}_{Pc}{}^2 \boldsymbol{\mu}_N{}^2 r_{12}/4\epsilon_0 h(\nu_N{}^2 - \nu_{Pc}{}^2) R_{12}{}^3\} \sin\alpha \sin^2\theta \qquad (6.20)$$

The ratio between the rotatory strengths of the symmetrical and anti-symmetrical components is only dependent on the geometrical parameters that are shown in Figure 6.4:

$$\begin{aligned}(R_1{}^A + R_2{}^A)/(R_1{}^B + R_2{}^B) = 2r_{12}\sin\alpha/(&-2R_{12}\sin^2(\alpha/2)\sin 2\theta \\ &+ r_{12}\sin\alpha\sin^2\theta) \qquad (6.21)\end{aligned}$$

By inserting the optimised parameters shown in Figure 6.4, a ratio that ranges from −2.1 to −3.94 is obtained for oxo- and sulfur-bridged substituents, respectively. This means that the two components have different signs, but the positively signed signal associated with the A symmetry component is at least twice as intense as the negatively signed signal due to the B symmetry component. In the context of the B component, the coupled oscillator and CD stealing mechanisms partially cancel each other, while the A component derives its intensity only from the "CD stealing" term. Hence, the CD signals of the Q and B1 bands in mono-(*R*)-binaphthylPc are expected to be dominated by the positively signed bands associated with the A symmetry component. For the *S* configuration, the signs of the observed signals are, of course, the opposite of those described for the *R* configuration. For compounds containing two and four binaphthyl units, the relative intensity simply doubles or quadruples compared with the mono-binaphthyl case, since the relative alignment of the Pc and binaphthyl chromophores remains unchanged.

Figure 6.9 contains an expanded view of the Q band regions of Figures 6.2 and 6.3. The Q band of tetra-binaphthyl substituted compounds has an inflection point in the MCD spectrum, which corresponds to the band centre of the corresponding absorption band. The presence of a derivative-shaped Faraday $\mathcal{A}_1$ term indicates that the excited state is orbitally degenerate. The Q band in the CD spectrum is similar in shape to that of the absorption spectrum. In contrast, the Q bands in the electronic absorption spectra of compounds containing two binaphthyl units at 686 nm are broader and have a shoulder on the shorter wavelength side at *ca.* 677 nm. A negative trough of intensity is observed in the MCD spectrum at 689 nm, while a peak of positive intensity lies at 678 nm, corresponding almost exactly to the centre of the broad band and the shoulder of intensity. This indicates that the excited states are non-degenerate and that the MCD spectrum is composed of Gaussian-shaped Faraday $\mathcal{B}_0$ terms. In addition, there is a narrowing of the Q band in the CD spectrum relative to the absorption band. These characteristics can also be rationalised as an extension of the explanation provided above.

Molecular orbital calculations were carried out for the compound shown in Figure 6.10, which has a π-conjugation system similar to the compound in Figure 6.3. In this region, two transitions are estimated to lie at 674 and 671 nm, which are assigned to the HOMO(a_u) → LUMO(b_{2g}) (B_{2u} excited state) and the HOMO(a_u) → LUMO + 1(b_{1g}) (B_{1u} excited state) transitions, respectively. As a result, the Q band at lower energy is polarised along the *y*-axis, while the band at higher energy is polarised along the *x*-axis. Comparison of this result with the overview of possible couplings in Figure 6.7 indicates that the coupling patterns shown at the top left and bottom right are associated with the *y*- and *x*-polarised transitions, respectively. The rotatory strengths of the former and latter transitions are given as twice those determined by equations (6.12), (6.14), (6.19) and (6.20), which describe values for a mono-binaphthyl Pc. The equation for the A symmetry component can be rewritten as:

$$R^{\mathrm{A}} = R_1{}^{\mathrm{A}} + R_2{}^{\mathrm{A}} = 2\rho \mathrm{r}_{12} \sin\alpha \qquad (6.22)$$

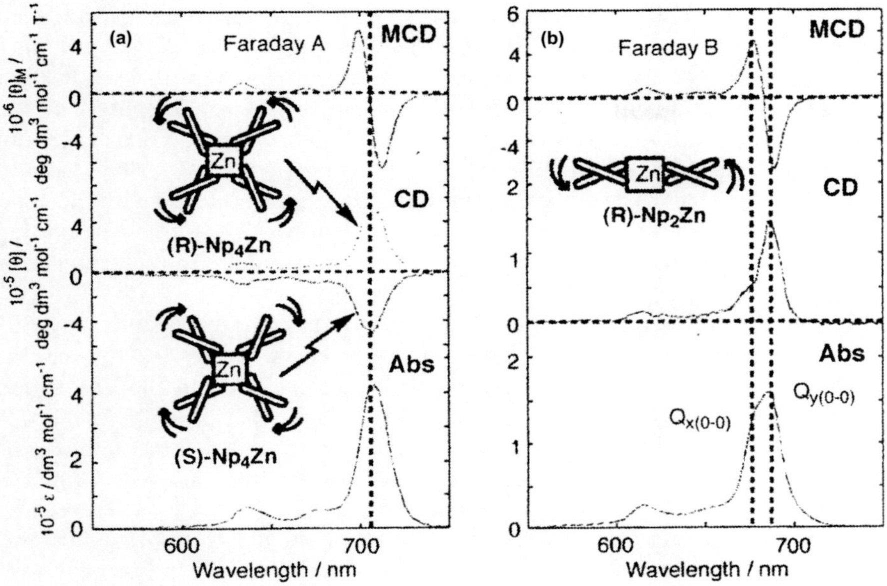

Figure 6.9 The electronic absorption, CD and MCD spectra of (*R*)- and (*S*)-Np_4Zn (left) and (*R*)-Np_2Zn (right) in the Q band region.[13]

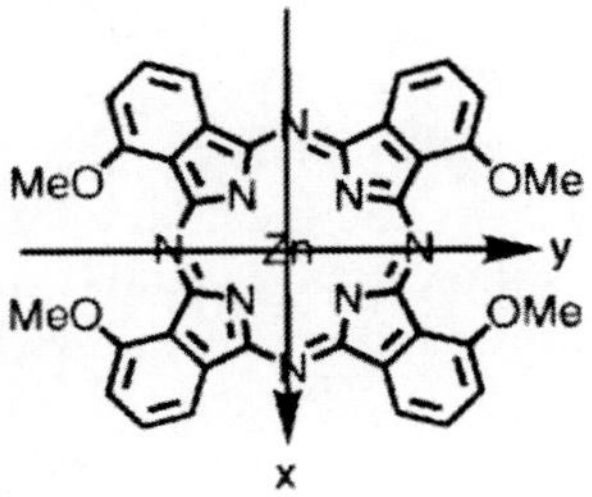

Figure 6.10 Q band polarisations.[6]

while that for the B symmetry component can be simplified into the following form:

$$R^B = R_1{}^B + R_2{}^B = \rho(-2R_{12}\sin^2(\alpha/2)\sin 2\theta + r_{12}\sin\alpha\sin^2\theta) \quad (6.23)$$

$$\text{where} \quad \rho = \nu_{Pc}\nu_N\boldsymbol{\mu}_{Pc}{}^2\boldsymbol{\mu}_N{}^2/2\epsilon_0 h(\nu_N{}^2 - \nu_{Pc}{}^2)R_{12}{}^3 \quad (6.24)$$

which is always positive in sign. The introduction of the θ, α, r_{12} and R_{12} values from Figure 6.4 into equations (6.22) and (6.23) produces values of 8.59ρ for

R_2^A and -4.11ρ for R^B. Since ρ is positive in sign, these values of rotational strength indicate that the ICD associated with the Q band of a Pc containing two binaphthyls is expressed as the sum of a large positive rotational strength to the longer wavelength side and a smaller negative rotational strength on the shorter wavelength side. In the experimental ICD spectrum, Figures 6.3 and 6.9, resolution is not observed, since the splitting of the Q(0,0) band is too small. For the reasons described above, the CD signal associated with the main Q band peak is significantly narrower than the bandwidth of the corresponding absorption band, where there is significant broadening due to the presence of hot bands.

The relative ICD intensity of the Q (*ca.* 550–750 nm) and B1 bands (*ca.* 300–450 nm) can also be readily explained. As observed in Figures 6.2 and 6.3, the apparent intensity (absorption coefficient) of the B1 band in the absorption spectrum is always much weaker than that of the Q band, while in the CD spectra the $[\theta]_M$ values in the B1 region become relatively more intense. As shown for example in equations (6.14) and (6.20), the induced rotational strength values (R^A, R^B) are proportional to $\nu_N\nu_{Pc}/(\nu_N^2 - \nu_{Pc}^2)$, where ν_N and ν_{Pc} are the frequencies of the absorption of naphthalene and phthalocyanine, respectively. As a result, the calculated rotational strength is inversely proportional to the difference between the ν_N^2 and ν_{Pc}^2 values. Since the most intense long-axis-polarised absorption band of naphthalene lies in the 220–225 nm region, and the Q and B1 bands of the phthalocyanine described in this section lie close to 700 and 370 nm, respectively, a relative intensification of the ICD intensity of the B1 band is anticipated. For example, the theoretical rotational strength ratio for TNpPcZn ($\nu_N = 45\,200$ cm^{-1}, ν_{Pc}(B1) $= 27\,200$ cm^{-1}, ν_{Pc}(Q) $= 14\,100$ cm^{-1}) is 1.12 (B1 band) : 0.38 (Q band), which is in close agreement with the experimentally observed ~3 (320–500 nm) : 1 (550–750 nm) ratio.

In summary, the sign, intensity and even the band shape of the CD spectra of binaphthyl-linked phthalocyanines can be readily accounted for in theoretical terms. Although a detailed explanation of the relevant theory has been provided in the literature,[6] no other example of this type of ICD system has been reported to date, but this approach has considerable potential if applied properly to other chromophores.

References

1. L. Pu, *Chem. Rev.*, 1998, **98**, 2405.
2. T. Nakano and Y. Okamoto, *Chem. Rev.*, 2001, **101**, 4013.
3. R. Grinter and S. F. Mason, *Trans. Faraday Soc.*, 1964, **60**, 274.
4. S. F. Mason, R. H. Seal and D. R. Roberts, *Tetrahedron*, 1974, **30**, 1671.
5. N. Harada and K. Nakanishi, *Circular Dichroism Spectroscopy – Exciton Coupling in Organic Stereochemistry*, University Science Books, New York, 1983.
6. N. Kobayashi, R. Higashi, B. C. Titeca, F. Lamote and A. Ceulemans, *J. Am. Chem. Soc.*, 1999, **121**, 12018.

7. I. Tinoco, *Adv. Chem. Phys.*, 1962, **4**, 113.
8. P. E. Shipper, *J. Am. Chem. Soc.*, 1979, **101**, 6826.
9. B. Nordén, A. Rodger and T. Dafforn, *Linear Dichroism & Circular Dichroism: A Texbook of Polarized-Light Spectroscopy*, Royal Society of Chemistry, Cambridge, 2010.
10. M.-C. Tsu and R. W. Woody, *J. Am. Chem. Soc.*, 1971, **93**, 5315.
11. T. Mizutani, T. Ema, T. Yoshida, Y. Kuroda and H. Ogoshi, *Inorg. Chem.*, 1993, **32**, 2072.
12. R. Grinter and S. F. Mason, *Trans. Faraday Soc.*, 1964, **60**, 274.
13. N. Kobayashi and A. Muranaka, *J. Synth. Org. Chem., Jpn.*, 2006, **64**, 735.

CHAPTER 7

Analysis of Chiral Systems by Theoretical Calculations

Optically active systems with rigid and well-defined geometries can be analysed on the basis of MO calculations. The spectra of systems with intrinsic chirality are best suited for this type of analysis. For example, the CD spectroscopy of aromatic compounds has been analysed in this manner since the 1980s. CD spectroscopy is highly sensitive to small conformational changes, so great care must be taken in analysing CD spectra using commercially available software. If X-ray data exist, those coordinates should be used to calculate CD solution spectra. In the context of porphyrinoids, for example, all conformational analyses reported to date have shown close agreement with the X-ray structures.[1]

In recent years, some researchers who are unfamiliar with the technique have calculated CD spectra on the basis of the predicted enantiomeric structures of low symmetry heterocyclic compounds. This should be avoided because near planar π-systems are often predicted inaccurately to have deformed low symmetry structures, resulting in unrealistic calculated spectra. CD spectral analyses of compounds for which no X-ray data are available, therefore, must be performed carefully. The most important factors are the distance and angles between chromophores. If the orientations of the edtms and mdtms are well defined structurally and their alignments can be readily estimated, the CD sign pattern can often be accurately predicted in a manner that can be used to determine the absolute configuration. Knowledge derived from the assignment of the electronic absorption spectra should be used wherever possible. For instance, the use of intense electronic absorption bands can be particularly helpful during CD studies. In the examples described below in Section 7.1, the CD and absorption spectra of twisted π-electron systems with intrinsic chirality were calculated using an older semi-empirical approach based on the π-electron self consistent field-configuration interaction-dipole velocity (SCF-CI-DV)

Circular Dichroism and Magnetic Circular Dichroism Spectroscopy for Organic Chemists
By Nagao Kobayashi, Atsuya Muranaka and John Mack

Published by the Royal Society of Chemistry, www.rsc.org

method.[2] More recent calculations carried out using modern time dependent-density functional theory (TD-DFT) techniques are described in section VII.2.

7.1 Semi-empirical Calculations

7.1.1 Absolute Stereochemistry of (+)-1,8a-Dihydro-3,8-Dimethylazulene[2]

1,8a-dihydro-3,8-dimethylazulene (**1**) was isolated from the cell culture of a liverwort, *Calypogeia granulate*, as a labile chiroptically active trinorsesquiterpenoid biosynthetic intermediate for 1,8-dimethylazulene.[3] Since the π-electron system of this compound consists of a twisted conjugated polyene chromophore, its absolute configuration had remained undetermined because of its extreme instability. Figure 7.1 contains the electronic absorption and CD spectra of compound **1**. A very intense CD band is observed, which is consistent with the presence of a strongly twisted, conjugated tetracene system. The absorption and CD spectra of a model compound, 1,8a-dihydroazulene (**2**), were calculated on the basis of the π-electron framework using the SCF-CI-DV molecular orbital (MO) method.[2]

In the context of the calculations, the absolute configuration was arbitrarily chosen to be (8a*R*) so that a comparison could be made between calculated and experimental spectra. The 1,8a-dihydroazulene structure is conformationally very rigid and the triene portion (carbon atoms 3a–4–5–6–7–8) of the

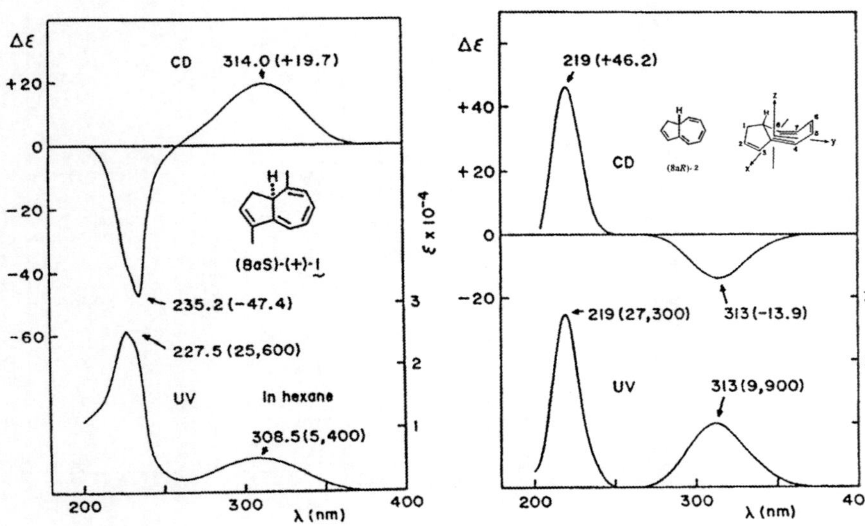

Figure 7.1 CD and UV spectra of naturally occurring (8a*S*)-(+)-1,8a-dihydro-3,8-dimethylazulene in hexane (left).[2b] CD and UV curves of (8a*R*)-1,8a-dihydroazulene calculated by the SCF-CI-DV MO method (right). [Reprinted with permission from *J. Am. Chem. Soc.*, 1985, **107**, 423. Copyright 1985, American Chemical Society.]

seven-membered ring is almost symmetrical when reflected through the *yz*-plane. In the absorption spectrum, two allowed $\pi \rightarrow \pi^*$ bands are predicted at 313 and 219 nm, in close agreement with the experimental data for **1**, Figure 7.1. The polarisation direction of the 313 nm band lies approximately along the *x*-axis, while that of the more intense band at 219 nm is aligned with the *y*-axis. Negatively and positively signed CD bands were predicted in the calculated spectrum with band centres that correspond closely to the absorption bands at 313 and 219 nm, Figure 7.1. The observed CD spectrum of **1** is almost a mirror image of that calculated for the (8a*R*)-**2** model compound. The absolute stereochemistry of (+)-**1** was, therefore, determined to be (8a*S*), as shown in structure **1**.

7.1.2 Analysis of Naphthalene–diene Derivatives

SCF-CI-DV calculations were also used to determine the absolute stereostructures of a series of halenaquinones, pentacyclic marine natural products isolated from *Xestospongia exigua* and *sapra* tropical sea sponges.[4] A halenaquinol dimethyl ether was converted successfully to a series of naphthalene–diene derivatives, which enabled a detailed structural analysis by CD spectroscopy. Figure 7.2 contains the absorption and CD spectra of the halenaquinol *trans*-methoxy diene derivative (3*R*,4*R*,12b*S*)-(–)-**3** in methanol. In the

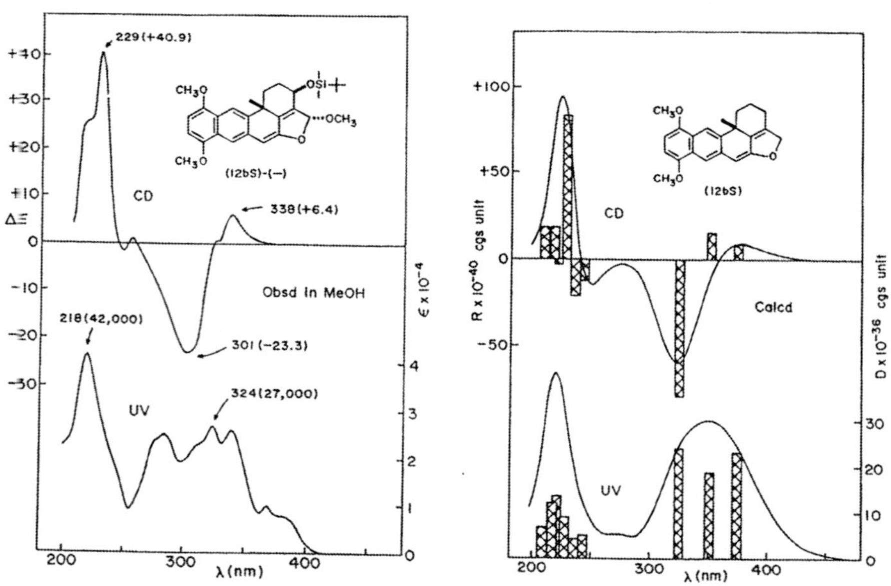

Figure 7.2 Observed CD and UV spectra of halenaquinol *trans*-methoxy diene derivative (3*R*,4*R*,12b*S*)-(–)-**3** in methanol (left). Rotational and dipole strengths of the transitions of a (12b*S*)-**4** model compound calculated by the π-electron SCF-CI-DV MO method (right).[4] [Reprinted with permission from *J. Am. Chem. Soc.*, 1989, **111**, 5668. Copyright 1989, American Chemical Society.]

absorption spectrum, an intense π → π* band with a complicated vibrational structure is observed at 324 nm (ε = 27 000), while a narrower band is observed at 218 nm (ε = 42 000). In the CD spectrum, three major intense bands are observed at 338 nm (Δε = +6.4), 301 nm (Δε = –23.3) and 229 nm (Δε = +40.9). The presence of a broad envelope of spectral bands out to 400 nm makes it obvious that the naphthalene–diene moiety, which is twisted by the angular methyl group at the 12b position, is the main chromophore of (–)-**3**. The (12b*S*)-**4** structure was adopted as a model compound for carrying out theoretical calculations of the CD spectra, since it contains the most relevant part of the π-electron system of the naphthalene–diene compound **3**. The chiral centres associated with the silyloxy group at the 3 position and the methoxy group at the 4 position are less significant where the observed CD intensity is concerned. The molecular geometry of the model compound was derived from a molecular mechanics calculation. The double bond and naphthalene chromophores form a clockwise helicity (dihedral angle of 5a–6–6a–7: +170°) in the (12b*S*) absolute configuration, while the conjugated diene moiety constitutes an anticlockwise helicity (dihedral angle of 3a–12c–5a–6: −167°). The helical sense of these two moieties is not changed, even if the D ring adopts a boat conformation, implying that the sense of the twist of the conjugated π-electron system is governed solely by the chirality of the angular methyl at the 12b position.

The calculated absorption spectrum, Figure 7.2, is comprised of two intense π → π* bands at 349 and 219 nm, which is in close agreement with the observed spectrum. The energies and relative intensities of the bands in the calculated CD spectra broadly resemble the experimental data, Figure 7.2. Nine major electronic transitions are included. The first and second bands in ascending energy terms are predicted to have weak positive rotational strengths and, therefore, probably correspond to the weak positively signed CD bands observed at longer wavelengths. The third transition is predicted to have an intense negative rotational strength. The negatively signed CD band, which is the lowest energy π → π* band with significant intensity, can be assigned to this transition. The sixth transition is predicted to result in strong positive rotational strength and probably corresponds to the intense positively signed CD band observed in the 220–230 nm region. Since the assignment of the component transitions to the observed CD signals is relatively straightforward, the absolute configuration of (–)-3 was determined to be 12b*S* on the basis of the SCF-CI-DV calculation.

7.2 TD-DFT Calculations

7.2.1 Correlation between CD Sign and Conformation in Optically Active Oxo(phthalocyaninato)vanadium(IV)

The CD spectroscopy of oxo(phthalocyaninato)vanadium(IV) provides an example of the use of modern TD-DFT calculations to predict the optical

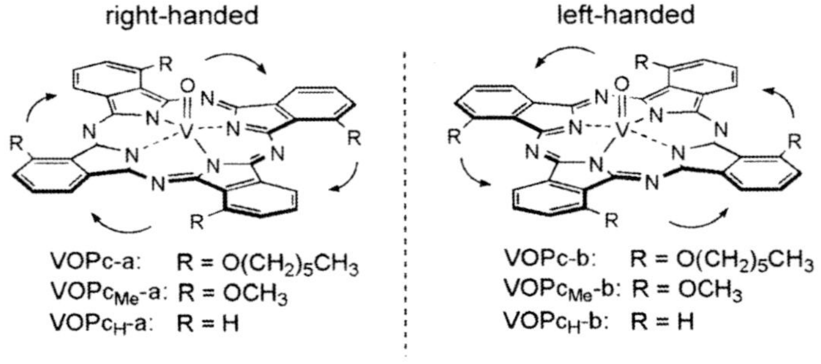

Figure 7.3 Structures and abbreviations of VOPc compounds.[5]

activity of a complex with inherent chirality. The Pc plane of oxovanadiumphthalocyanine (VOPc) is slightly domed, since the oxygen atom is an axial ligand. As a result, when VO is inserted into a Pc ligand in which peripheral substituent groups have been introduced in either a left- or right-handed manner, the resulting VOPc complex contains a mixture of two enantiomers, Figure 7.3. Pure enantiomers can be obtained after separation using optically active columns.[5] The CD spectra of the enantiomers contain ICD bands in the wavelength regions where the Pc chromophore absorbs strongly. The first and second fractions to be eluted by high-performance liquid chromatography (HPLC) were designated as VOPc1 and VOPc2, respectively. The first fraction has a negatively signed CD band in the Q band region and a plus-to-minus and a plus-to-minus pattern in ascending energy in the B1 band region, Figure 7.4.

VOPc1 and VOPc2 consist of an achiral Pc with four achiral hexyloxy substituents. The peripheral alkoxy substituent has an edtm that is so small that any ICD contribution due to the coupled dipole and "CD stealing" mechanisms, see Section 1.2.2, is negligible at best, making it feasible to analyse the inherent chirality of the system. In this kind of system, the CD analysis method is limited to direct calculations using the Rosenfeld equation[6] based on the complete molecular wavefunctions. TD-DFT calculations were carried out for C_4 symmetry VOPc model complexes containing four methoxy groups. The groundstate geometry of the right-handed isomer, $VOPc_{Me}$-a, was calculated with the symmetry restricted to C_4 using the B3LYP functional with 6–31G* basis sets.[7] The geometry of the alkoxy groups in the initial structure was based on that reported in the crystal structure of a NiPc containing alkoxy substituents at the same peripheral positions. There is a significant doming distortion in the optimised structure of the Pc ligand, similar to that observed for unsubstituted VOPc. The geometric parameters for $VOPc_{Me}$-a are almost identical to those of the crystal structure.

The calculated absorption and CD spectra are shown in Figure 7.5. The calculations predict three absorption bands at 646, 413 and 360 nm, which can be assigned to the bands observed at 733, 434 and 353 nm, respectively. The lowest-energy band is associated with a doubly degenerate transition and corresponds to the negative Gaussian-type CD signal observed for VOPc1.

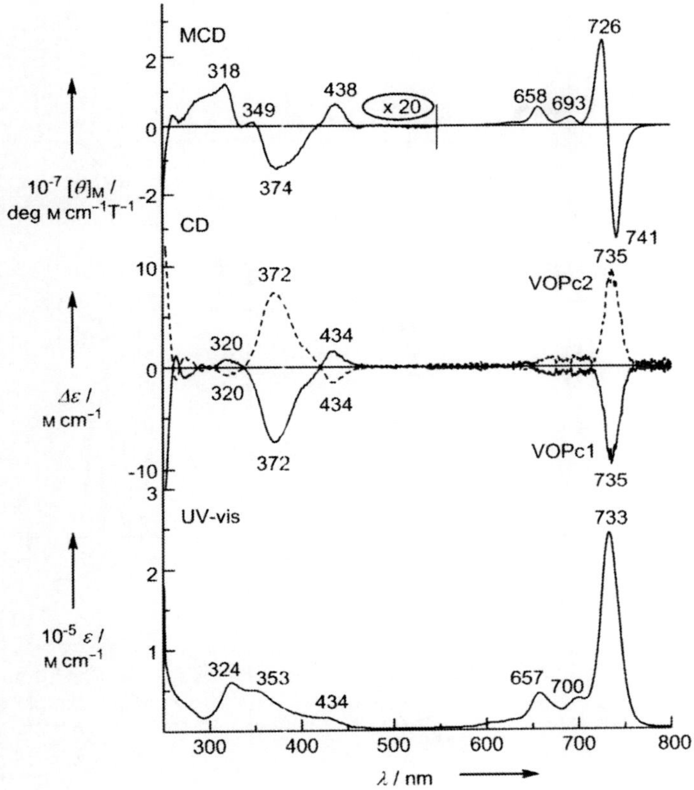

Figure 7.4 MCD, CD and electronic absorption spectra of VOPc.[5]

Positive and negative CD signals are predicted for the 413 and 355 nm transitions respectively, which is consistent with the observed positive and negative CD signals at 434 and 372 nm. The first eluted enantiomer was, therefore, concluded to have a right-handed geometry (VOPc-a).

7.2.2 Optically Active Porphyrin Dimers

A doubly linked porphyrin dimer was synthesised by Osuka and coworkers, Figure 7.6, and its enantiomers were separated by HPLC with an optically active column.[8] As shown in Figure 7.7, the absorption spectrum of the fused dimer differs markedly from that of the nickel tetraphenylporphyrin monomer. An intensely red shifted band is observed at 736 nm with a shoulder at *ca.* 820 nm. The MCD spectrum differs markedly from those typically observed for metalloporphyrins, since derivative-shaped Faraday $\mathcal{A}_1$ terms are not observed due to the lack of a three-fold or higher axis of symmetry.[9] The CD spectrum is not consistent with exciton coupling, because Gaussian-shaped bands are

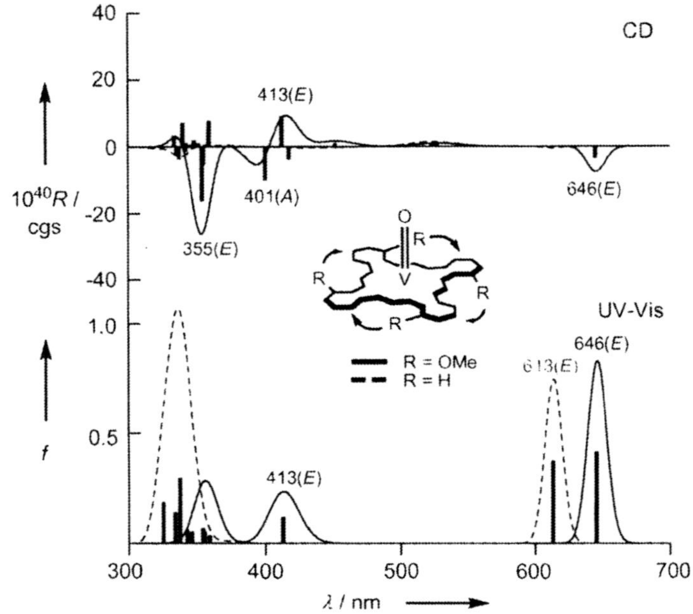

Figure 7.5 Simulated electronic absorption and CD spectra of $VOPc_{Me}$–a. The dashed lines denote simulated spectra of $VOPc_{H}$–a in which four methoxy groups are replaced by hydrogen atoms. Gaussian bands with half-bandwidth of 400 cm^{-1} (Q band region) and 1500 cm^{-1} (B1 band region) were used.[5]

observed at wavelengths longer than *ca.* 700 nm. Exciton coupling theory is not relevant in this context, given that the constituent chromophores are not independent. The CD analysis can, therefore, be carried out based on direct calculation from the Rosenfeld equation[6] using the complete molecular wavefunctions. Geometry optimisation calculations were performed at the DFT level of theory for a right-handed helical *bis*porphyrin with C_2 symmetry to elucidate the relationship between the observed spectral features and the helical structure of the *bis*porphyrin. The optimised geometry was predicted to adopt a ruffled conformation, very similar to that obtained by X-ray crystallography.

The calculated absorption and CD spectra exhibit very close agreement with the experimental data, although the calculations overestimate the excitation energies and intensities significantly. Selected results are summarised in Table 7.1. Owing to the C_2 symmetry, transitions to A symmetry excited states are *z*-axis polarised, while those to B symmetry excited states are polarised in the *xy*-plane. The calculations predict two absorption bands, at 666 and 745 nm, which can be assigned, respectively, to the absorption bands observed at 735 nm and a shoulder observed at *ca.* 820 nm, based on their positive and negative CD intensities. The helical sense of the enantiomer was determined to be the right-handed geometry shown in Figure 7.6. The three lowest energy

Table 7.1 Selected TD-DFT results for a model *bis*porphyrin with a right-handed helical conformation.[9]

No.	*λ*/nm	*Symmetry*	*Oscillator strength*	*Rotatory strength*[a]	*Composition* / wt %
1	744.82	B	0.1427	–27.14	H→L(49.8); H–1→L(36.9); H–1→L+1(4.3)
2	666.14	B	0.3967	139.68	H–1→L(47.6); H→L(27.9); H→L+1(10.5); H–1→L+1(5.0)
3	595.97	A	0.0003	–0.47	H–2→L(58.4); H→L+2(39.7)
10	554.07	B	0.1586	314.02	H→L+1(49.7); H–4→L(32.5); H–1→L(1.9)
12	5.6.24	B	0.0737	190.61	H–4→L(37.2); H→L+1(21.2); H–6→L(16.5); H–1→L(3.3)
13	504.81	A	0.1374	–141.95	H→L+2(29.9); H–2→L(17.1); H–1→L+3(12.3); H–5→L(9.9); H–7→L(7.6); H–1→L+2(5.6); H→L;3(3.3)
15	498.07	B	0.3358	61.38	H–1→L+1(39.1); H–8→L(21.6); H–6→L(14.5); H–4→L(5.2)
24	443.51	B	0.8973	–64.03	H–2→L+2(51.7); H–4→L+1(7.8); H–6→L+1(6.2); H–3→L+3
34	391.73	A	0.1822	–457.16	H–7→L+1(42.0); H→L+3(7.0); H–2→L+1(5.8); H–3→L+1(5.7); H–1→L+2(5.0)
35	384.14	A	0.5732	–154.24	H–4→L+3(23.6); H–3→L+1(18.0); H–1→L+3(15.0); H–7→ L+1(11.3); H–6→L+2(6.2)
39	374.78	B	0.0358	221.45	H–5→L+2(23.7); H–2→L+3(21.7); H–6→L+1(16.1); H–3→L+2(9.4); H–10→L(7.1); H→L+6(3.0)

[a]cgs (10^{-40} erg esu cm G^{-1}).

bands can be assigned to the Q transition, Table 7.1. The two lowest energy bands are optically allowed and are polarised perpendicular to the *z*-axis, while the third transition is optically forbidden and is polarised along the *z*-axis.

7.2.3 Absolute Configuration and Chiroptical Properties of Three-layered [3,3]Paracyclophane

Three-layered [3,3]paracyclophane ([3,3]PCP), **1**, was resolved into two enantiomers, (*R*)-(–)-**1** and (*S*)-(+)-**1**, Figure 7.8. The absolute configurations were determined from a comparison of the experimental chiroptical properties and the results of TD-DFT calculations.[10] In order to gain an insight into the structure in solution, a conformational analysis of a three-layered [3,3]PCP with *R*-configuration ((*R*)-**1**) was carried out using the B3LYP functional with TZVP basis sets and the results were compared to the X-ray structure of (–)-**1**. Although the ^{1}H nuclear magnetic resonance (NMR) data appeared to be consistent with D_2 symmetry, seven stable conformations are predicted, Figure 7.9, which can be classified into three groups on the basis of their

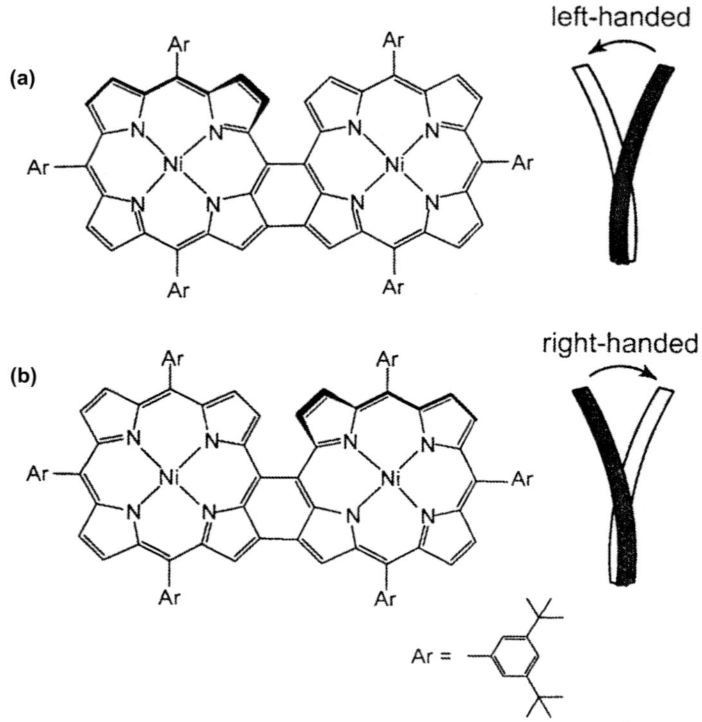

Figure 7.6 a) Doubly linked *bis*porphyrin and b) origin of chirality based on the crystal structure.[11]

symmetry (D_2, C_2 and C_1). The X-ray structure has C_2-a symmetry and adopts a boat–boat conformation. The differences in geometry between the conformations lie predominantly in the relative orientation of the carbon bridges. Thus, in all of the conformers the benzene rings are nearly stacked, as is observed in the crystal structure. The interplanar distances of these optimised structures (3.39–3.41 Å) are substantially longer than those of the crystal structure (3.260 and 3.284 Å). The geometrical parameters of the lowest-energy C_2-a symmetry conformation are essentially identical to those of the crystal structure.

Figure 7.10 contains the calculated absorption and CD spectra for C_2-a. Although not shown, the six other conformers in Figure 7.9 are predicted to have similar spectra to this, including the intense negatively signed CD signal at *ca*. 230 nm. This demonstrates that the geometry of the carbon bridge has a marginal role in determining the optical properties. Figure 7.10 contains experimental spectra for (+)-**1** and (–)-**1**, together with a Boltzmann-averaged spectrum for the seven conformers predicted for (*R*)-**1**. It appears reasonable to assume on this basis that the positively and negatively signed CD signals observed for the highest energy band at the instrumental limit and the most

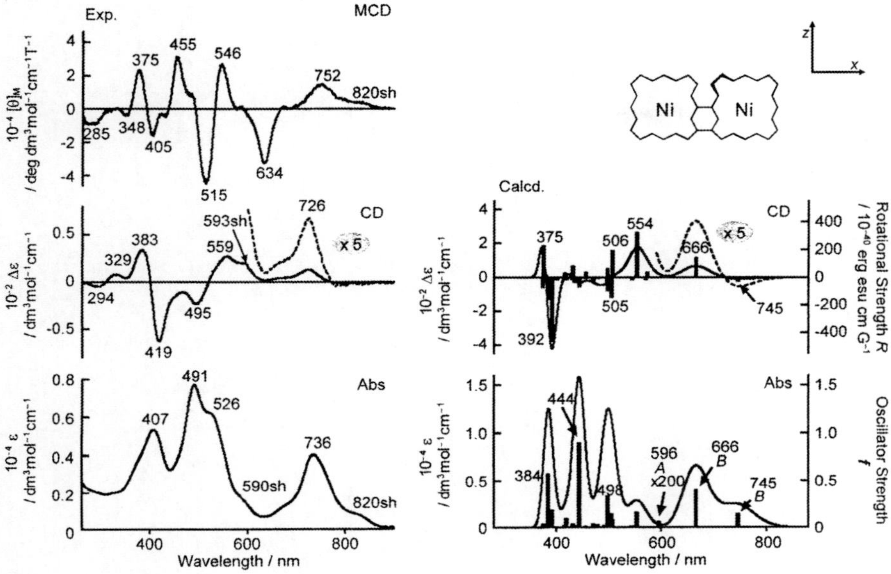

Figure 7.7 Experimental MCD, CD and electronic absorption spectra of one of the enantiomers in Figure 7.6.[9] Calculated CD and absorption spectra obtained as sums of a Gaussian-type function with a half bandwidth of 1300 cm^{-1} (left scale) (right). The CD and absorption intensities are provided as rotational strengths, *R*, and oscillator strengths, *f*, respectively (right scale). The absorption intensity at 596 nm is magnified 200 times. *A* and *B* are symmetry labels.

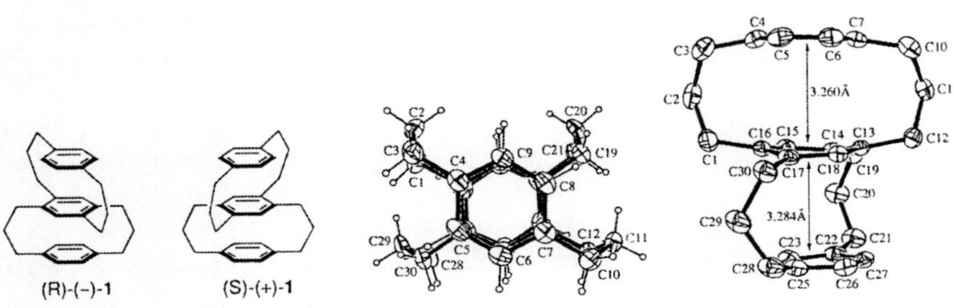

Figure 7.8 Structures of (*R*)-(−)-**1** and (*S*)-(+)-**1** and molecular structures of (−)-**1** (top and side views).[10]

intense band in the 200–250 nm region of the (−)-**1** spectrum are associated with an (*R*)-**1** structure. In order to confirm this relationship further, the specific rotation at the sodium D line ($[\alpha]_D$) was calculated for each conformer of (*R*)-**1** at the same level of theory as the CD calculations. All of the conformers have negative values from −124.4 to −221.8°, which is in agreement with the

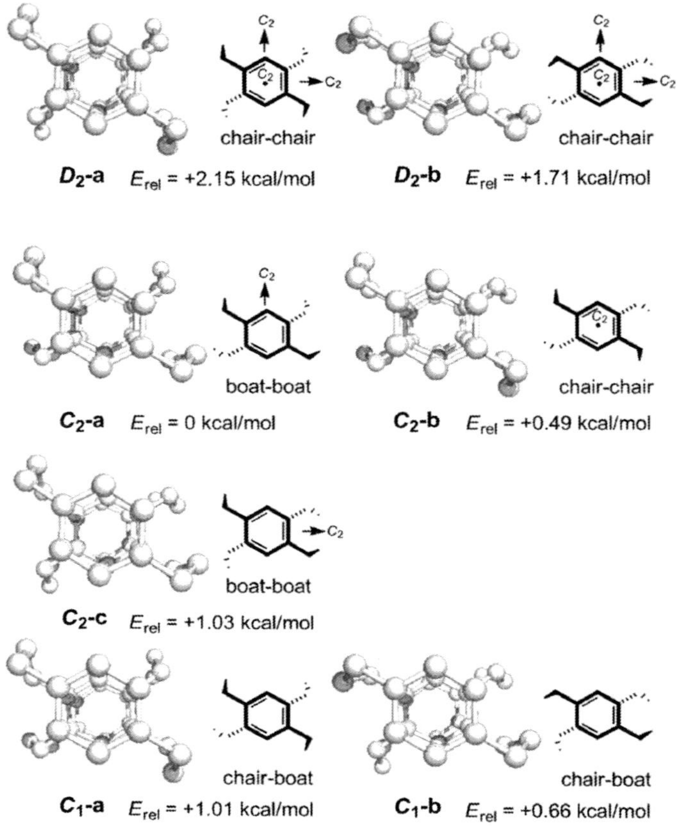

Figure 7.9 B3LYP/TZVP structures of three-layered [3,3]PCP, (*R*)-**1**. The relative energies (E_{rel}) are based on total energies corrected with the zero-point vibrational energies.[10]

experimental $[\alpha]_D$ value of −123° in CH_2Cl_2 for (−)-**1**. The TD-DFT calculations predict that $[\alpha]_D$ is negative for (*R*)-**1** and and positive for (*S*)-**1**, so the absolute configuration of [3,3]PCP was determined to be (*R*)-(−)/(*S*)-(+).

Given that the contribution of the carbon bridge to the observed CD signal was found to be relatively small, the chiroptical properties of (*R*)-**1** were investigated to determine whether they can be reproduced using a *p*-xylene–durene–*p*-xylene system model compound with an (*R*)-configuration. Almost identical absorption and CD spectra were obtained, which were similar in shape and intensity to those observed in Figure 7.10. Close agreement was obtained with the experimental data when spectra were calculated for the model system with the inter-ring separations and angles of the three benzene derivatives kept close to those of the X-ray structure of (−)-**1**. It is clear, based on these calculations, therefore, that the twisted structure and the deformation of the aromatic rings have very little effect on the spectral properties.

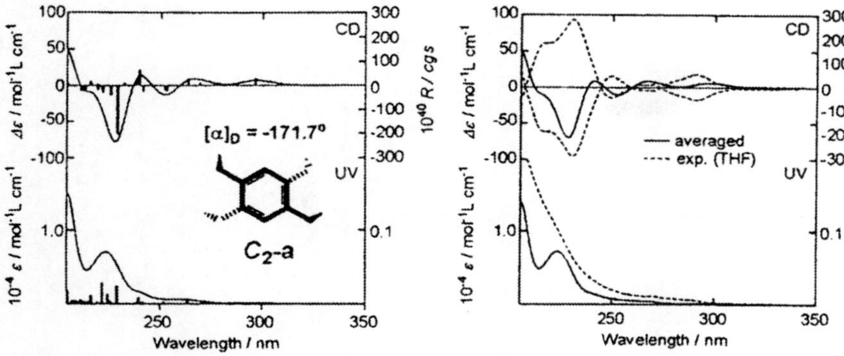

Figure 7.10 Calculated CD and absorption spectra for the C_2–a conformer of (*R*)-**1**; the calculated optical rotation value (B3LYP/TZVP) is also shown. Gaussian bands with a half-bandwidth of 2000 cm^{-1} were used to produce the spectra. $[\alpha]_D$ values are in deg [dm $(g/cm^3)]^{-1}$ (left). Boltzmann-averaged spectra for all seven of the predicted conformers for (*R*)-**1** are reasonably similar to the experimental spectra for for (+)-**1** and (–)-**1** (right).[10]

References

1. N. Kobayashi, in *Handbook of Porphyrin Science*, Vol. 7, ed. K. M. Kadish, K. M. Smith and R. Guilard, World Scientific, Singapore, 2010, p. 147.
2. (a) N. Harada in *Circular Dichroism*, ed. N. Berova, K. Nakanishi and R. W. Woody, Wiley–VCH, New York, 2000, p. 431; (b) N. Harada, J. Kohori, H. Uda, K. Nakanishi and R. Takeda, *J. Am. Chem. Soc.*, 1985, **107**, 423.
3. R. Tanaka and K. Katoh, *J. Am. Chem. Soc.*, 1983, **105**, 4056.
4. H. Harada, H. Uda, M. Kobayashi, N. Shimizu and I. Kitagawa, *J. Am. Chem. Soc.*, 1989, **111**, 5668.
5. N. Kobayashi, F. Narita, K. Ishii and A. Muranaka, *Chem. Eur. J.*, 2009, **15**, 10173.
6. V. L. Rosenfeld, *Z. Phys.*, 1928, **52**, 161.
7. F. Jensen, *Introduction to Computational Chemistry*, Wiley, New York, 2nd edn, 2007.
8. A. Tsuda, H. Furuta and A. Osuka, *J. Am. Chem. Soc.*, 2001, **123**, 10304.
9. A. Muranaka, Y. Asano, A. Tsuda, A. Osuka and N. Kobayashi, *ChemPhysChem.*, 2006, **7**, 1235.
10. A. Muranaka, M. Shibahara, M. Watanabe, T. Matsumoto, T. Shinmyozu and N. Kobayashi, *J. Org. Chem.*, 2008, **73**, 9125.
11. N. Kobayashi and T. Fukuda, *Bull. Chem. Soc. Jpn.*, 2009, **82**, 631.

CHAPTER 8

Circular Dichroism of Biomolecules

Despite the fact that almost 50 years have elapsed since circular dichroism (CD) spectroscopy was first applied to biomolecules, many new CD spectra are published each year for natural systems. The CD spectra of biomolecules continue to be reported routinely by many laboratories because of the simplicity of the measurement technique.[1] The same solutions and cuvettes can often be used to carry out both the electronic absorption and CD measurements. The key biomolecules that have been studied in this manner, such as proteins and nucleic acids, are generally very large, so detailed theoretical analyses tend to be problematic. The spectral data, therefore, can often only be analysed in a qualitative manner using empirically derived relationships.[1] Proteins are linear polymers of well-defined sequences of twenty L-amino acids linked *via* peptide bonds, which are formed between the carboxyl group of one amino acid and the animo group of another with the elimination of a water molecule. Most of the constitutent amino acids are chiral and proteins often adopt specific secondary and tertiary structures, which optimise their biological activity. Polynucleotides such as deoxyribonucleic acid (DNA) and ribonucleic acid (RNA) are biopolymers composed of achiral nucleotide monomers, which are covalently bonded in a chain based on backbones comprised of sugars and phosphate groups joined by ester bonds. DNA consists of two chains of polynucleotides; each chain forms a helical spiral, resulting in significant CD intensity.

CD spectra are often used to study changes in the folding of protein molecules in solution,[1] Figure 8.1, and this can sometimes be used to derive quantitative information in the context, for example, of metal binding by cysteine residues. The technique has also been used to probe the interaction of proteins with other molecules, often on the basis of the presence of induced circular dichroism (ICD) signals when these molecules are achiral, Figure 8.2. Key advantages relative to nuclear magnetic resonance (NMR) spectroscopy are

Circular Dichroism and Magnetic Circular Dichroism Spectroscopy for Organic Chemists
By Nagao Kobayashi, Atsuya Muranaka and John Mack

Published by the Royal Society of Chemistry, www.rsc.org

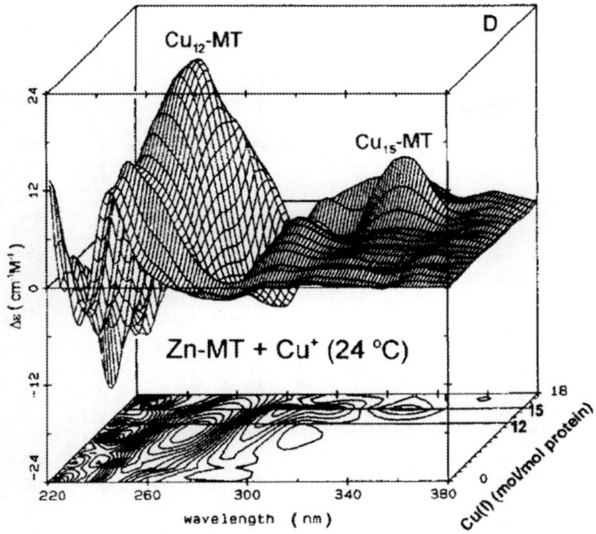

Figure 8.1 Three-dimensional projection plot of CD spectra recorded during the replacement of Zn^{2+} in a solution of Zn–metallothionein 2 (Zn–MT) at 24 °C.[5] Note the local maximum at 300 nm with 9.0 Cu^+ and at 335 nm with 15.0 Cu^+. The formation of new species at these ratios is also highlighted by the grid lines added to the contour plot. [Reprinted with permission from *Eur. J. Biochem.*, 1995, **227**, 226. Copyright 1995, FEBS.]

that measurements can be made at significantly lower concentrations and on a far shorter intrinsic timescale because the incident electromagnetic radiation consists of UV region light rather than radio waves.

8.1 Protein Chromophores and Electronic Absorption Spectroscopy

Although the peptide unit is planar and rigid, it still has two degrees of freedom in its alignment relative to the rest of the protein chain, Figure 8.3. A variety of different intramolecular hydrogen bonding arrangements between the C=O of one peptide unit and the N–H of another enable the formation of several different structural motifs such as the α-helix and β-sheet, Figures 8.4 and 8.5. Of the twenty amino acids, only phenylalanine, tyrosine and tryptophan have transitions in the 240–320 nm region (the near UV region): phenylalanine has one transition with ε_{max} (258 nm) ~190 mol^{-1} dm^3 cm^{-1}, tyrosine has one transition with ε_{max} (274 nm) ~1400 mol^{-1} dm^3 cm^{-1}, while the isoindole moiety of tryptophan has two or more transitions in the 240–290 nm region with a total maximum ε_{max} (279 nm) ~5000 mol^{-1} dm^3 cm^{-1}. Although the ε_{max} value of tryptophan is by far the most intense in this region, it is not necessarily the dominant factor in protein spectroscopy, since most proteins have relatively few

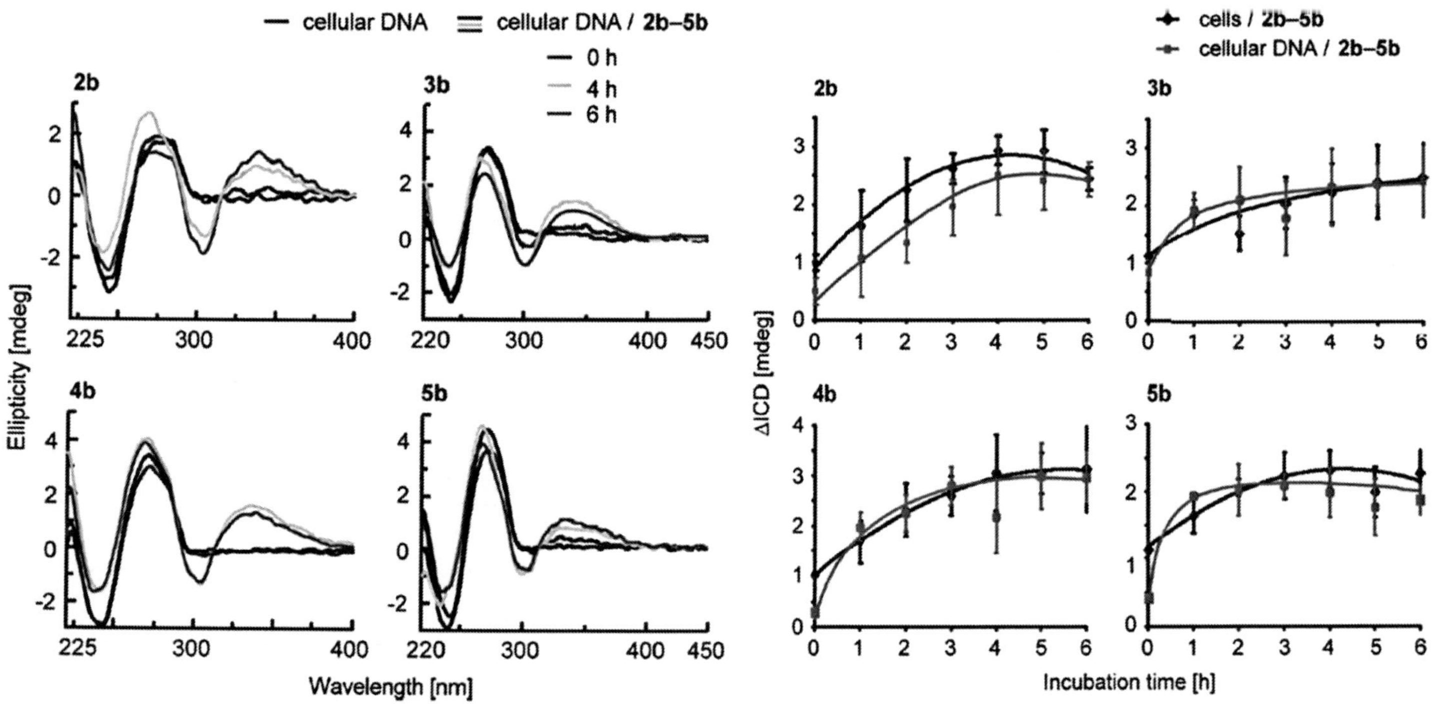

Figure 8.2 The observed ICD intensity was used to identify DNA as the cellular target of several novel anticancer agents based on the highly cytotoxic natural antibiotic CC-1065.[6] Ellipticity of cellular DNA after exposure of cells to the *seco*-drugs, **2b** – **5b**, for the time indicated (left). Induced CD (ΔICD = Ellipticity($\lambda = 335$ nm) – Ellipticity($\lambda = 305$ nm)) after exposure of cells to the *seco*-drugs **2b** – **5b**. All data points reflect mean values ± standard deviation (sd) (right). [Reprinted with permission from *J. Am. Chem. Soc.*, 2009, **131**, 13031. Copyright 2009, American Chemical Society.]

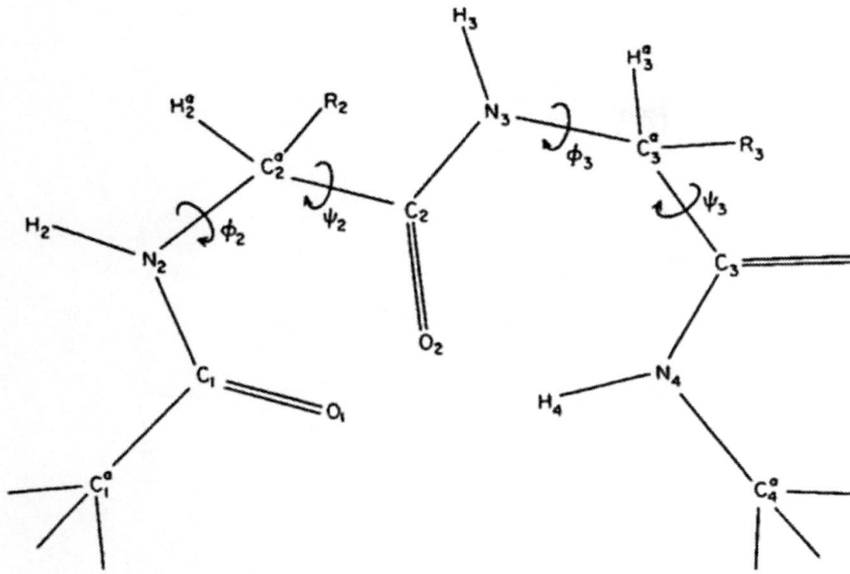

Figure 8.3 The two rotational degrees of freedom in each peptide unit, designated by the dihedral angles φ and Ψ, provide a peptide chain with its flexibility.[7] [Reprinted with permission from *J. Mol. Biol.*, 1976, **107**, 85. Copyright 1976, Elsevier.]

tryptophans compared to other aromatic residues. The far UV region (<250 nm) tends to be dominated by transitions of the peptide backbone of the protein. The peptide chromophore has non-bonding electrons on the oxygen and nitrogen atoms, π-electrons which are delocalised to a certain extent over the carbon, oxygen and nitrogen atoms, and σ-bonding electrons. Given the lower symmetry of the amide chromophore, the n → π* transition is not strictly forbidden in electric dipole terms. Thus, weak n → π* bands associated with the carbonyl oxygen atom lie in the 210–230 nm region ($\varepsilon \sim 100\ mol^{-1}\ dm^3\ cm^{-1}$) with the wavelength being determined by the extent of hydrogen bonding of the oxygen lone pairs. In aprotic solvents, the n → π* band lies near 230 nm, while in aqueous media the band lies close to 210 nm. The n → π* transition is magnetically allowed, with an mdtm of ~1 μ_B, directed approximately along the carbonyl bond, Figure 8.3. In secondary amides, the π → π* bands at *ca.* 185–190 nm (often termed the NV1 band, $\varepsilon \sim 7000\ mol^{-1}\ dm^3\ cm^{-1}$) and at *ca.* 140 nm (often termed the NV2 band) are electric dipole allowed and the molecular orbitals (MOs) involved are primarily associated with the carbonyl π-bond, Figure 8.6, but are also influenced to a certain extent by the lone pair orbitals of the nitrogen atom.[2] The transition dipole moments of the two π → π* transitions of secondary amides are approximately parallel (NV1) and perpendicular (NV2) to the C–N bond direction, Figure 8.3.

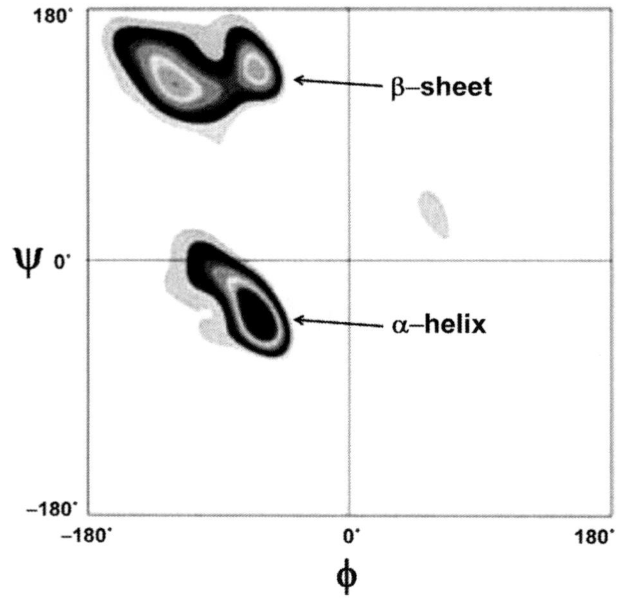

Figure 8.4 A Ramachandran plot shows the somewhat arbitrary boundaries between the values of φ and Ψ that result in the formation of β-strands (β) α-helices (α) and random coils (the remainder of the diagram).[8] [Reprinted with permission from *Acta Cryst.*, 2002, **D58**, 768. Copyright 2002, International Union of Crystallography.]

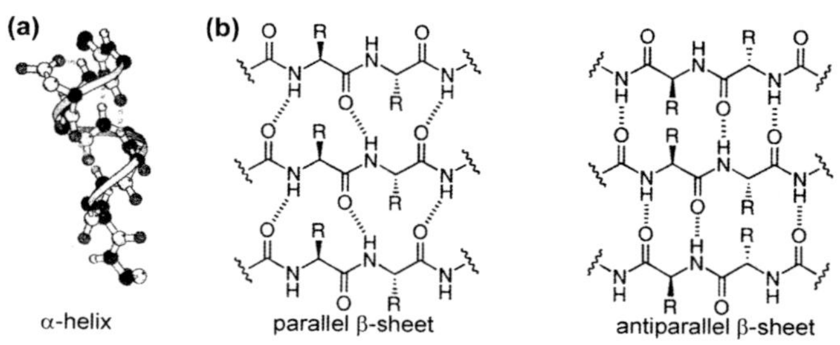

Figure 8.5 (a) α-Helix (left) and (b) β-sheet (right) protein secondary structures. Dotted lines indicate hydrogen bonds.[9,10] [Reprinted with permission from *J. Mol. Biol.*, 1999, **287**, 127 and *Tetrahedron*, 2002, **58**, 727. Copyright 1999 and 2002, Elsevier.]

Chromophores with far-UV transitions include the aromatic side chains, disulfide cysteines, arginine, asparagine, aspartic acid, glutamine, glutamic acid and histidine. Although some of the transitions in these chromophores are often more strongly allowed than the peptide $\pi \rightarrow \pi^*$ transitions, the side chain transitions are generally impossible to detect, since the peptide chromophores are in large excess. However, the presence of these side chain transitions can be

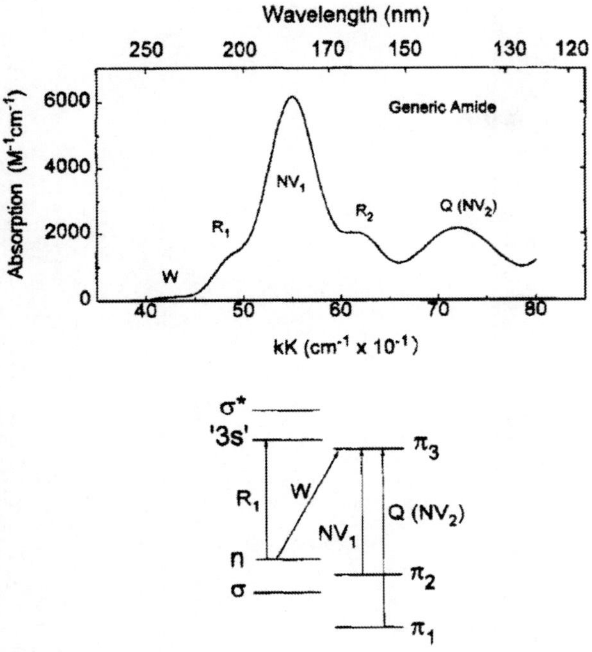

Figure 8.6 A generic amide vapour absorption spectrum (top) assembled as a composite based on various amide spectra. The π-MOs labelled here as π_1, π_2 and π_3 are sometimes also referred to as the π_+, π_o and π-MOs, respectively. The assignment of the bands to specific one-electron transitions is shown with arrows (bottom). The R_1 and R_2 bands have been assigned as Rydberg transitions.[2] [Reprinted with permission from *J. Am. Chem. Soc.*, 1995, **117**, 7974. Copyright 1995, American Chemical Society.]

sufficient to make the accurate empirical determination of the percentage of α-helical content problematic.

8.2 CD Spectroscopy of Peptides

The electric dipole allowed $\pi \rightarrow \pi^*$ transitions of protein molecules can undergo coupled-oscillator interactions with other chromophores in the molecule. The interaction among peptide groups provides the most striking example of this type of coupling. Distinctive CD spectra are observed on the basis of the type of peptide structure that is adopted, *i.e.* α-helix, β-sheet, β-turn or random coil, Figure 8.7. Given that the CD spectra of natural proteins are usually expressed as the sum of the percentages of these four component spectra, a variety of methods have been developed to determine the percentage of each structure.[1]

8.2.1 α-Helix CD

The α-helix is the dominant secondary structure in many proteins and on average accounts for about one-third of the residues in globular proteins. The

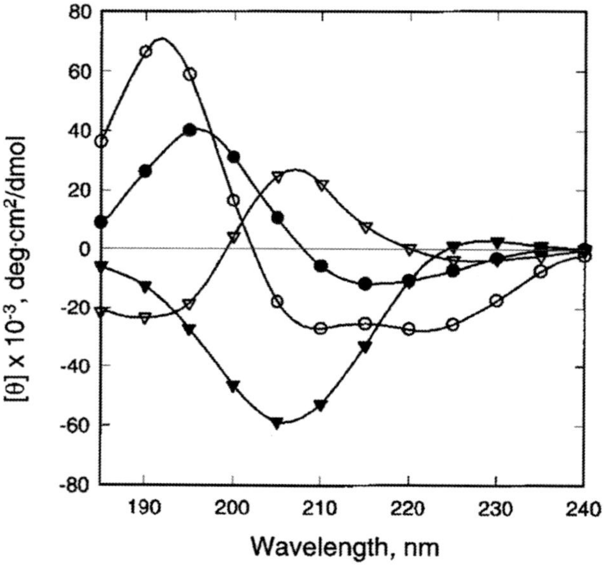

Figure 8.7 Circular dichroism spectra of polypeptides in the α-helix (open circles), β-sheet (solid circles), β-turn (open triangles) and random coil (solid triangles) conformations. Spectra like these can be used to fit CD spectra of proteins to determine the fractions of the structure, which are comprised of each secondary structure type.[3c] [Reprinted with permission from *Anal. Biochem.*, 1996, **235**, 1. Copyright 1996, Elsevier.]

spectroscopic characterisation has been facilitated greatly by the fact that many synthetic homopeptides, such as poly(L-Lys), adopt this conformation under appropriate conditions.[3] The CD spectra are characterised by a negative envelope of intensity with separate maxima of similar magnitude at 222 nm (n $\rightarrow$ π^*) and 208 nm ($\pi \rightarrow \pi^*$) and an intense positive peak in the 185–190 nm ($\pi \rightarrow \pi^*$, NV1 transition). The $[\theta]_M$ values at 222 and 190 nm are about $-40\,000$ to $-42\,000$ and 90 000 to 93 000, respectively, for homopolypeptides. These values vary depending on the degree of polymerisation. Generally, the larger the polymer the more intense the signal is. If the assumption is made that these are the $[\theta]_M$ values for a 100% α-helix, a rough α-helix content can be estimated using the experimental $[\theta]_M$ values at these wavelengths. For oligopeptides consisting of a small number of residues, the estimation of the helix content must be carried out carefully, since about four residues will not contribute to this CD intensity because one α-turn requires the presence of four amino acid residues and there are "end effects" associated with unanchored hydrogen-bonding groups.

8.2.2 β-Sheet and random coil CD

β-Sheets are formed by internal hydrogen bonds between either parallel or antiparallel β strands and are much more inflexible than an α-helix. Their analysis by CD spectroscopy is complicated by the fact that they are less structurally defined than α-helices and tend to be insoluble in solvents with a high level of transmission in the UV region. A negatively signed band is typically observed at *ca.* 215 nm, while a positive band of comparable intensity lies at *ca.* 195 nm, see Figure 8.6.[3] Unfortunately the empirical analysis of protein spectra is complicated by the fact that the positive and negative intensity maxima typically observed for random coil structures, see Figure 8.4, also lie at similar wavelengths.

8.2.3 β-Turn CD

The term β-turn is used to describe all of the turns that occur within a protein structure other than those involved in forming a β-sheet from a single strand. Typically about 25% of residues within a protein fall into this category. A weak positively signed n → π* band that is observed in the CD spectrum at *ca.* 225 nm is followed at higher energy by positive and negative π → π* bands at *ca.* 203 and 185 nm, respectively. Infrared (IR) spectroscopy has been found to be superior to CD spectroscopy for determining the proportion of β-turn structures, so the two techniques are often used in a complementary manner to study the secondary structure of proteins.[4]

References

1. (a) N. Sreerama and R. W. Woody, *Anal. Biochem.*, 1993, **209**, 32; (b) N. Sreerama and R. W. Woody, in *Circular Dichroism*, ed. N. Berova, K. Nakanishi and R. W. Woody, Wiley–VCH, New York, 2000, p. 601.
2. L. B. Clark, *J. Am. Chem. Soc.*, 1995, **117**, 7974.
3. (a) N. Greenfield and G. D. Fasman, *Biochemistry*, 1969, **8**, 4108; (b) S. Brahms and J. Brahms, *J. Mol. Biol.*, 1980, **138**, 149.
4. R. W. Sarver, Jr. and W. C. Krueger, *Anal. Biochem*, 1991, **199**, 61.
5. A. Presta, A. R. Green, A. Zelazowski and M. J. Stillman, *Eur. J. Biochem.*, 1995, **227**, 226.
6. L. F. Tietze, B. Krewer, F. Major and I. Schuberth, *J. Am. Chem. Soc.*, 2009, **131**, 13031.
7. G. N. Ramachandran and A. K. Mitra, *J. Mol. Biol.*, 1976, **107**, 85.
8. S. Hovmöller, T. Zhou and T. Ohlson, *Acta Cryst.*, 2002, **D58**, 768.
9. S. Penel, E. Hughes and A. J. Doig, *J. Mol. Biol.*, 1999, **287**, 127.
10. J. S. Nowick, E. M. Smith, J. W. Ziller and A. J. Shaka, *Tetrahedron*, 2002, **58**, 727.

CHAPTER 9

Analysis of MCD Spectra

In this chapter, typical examples of the analysis of the three Faraday terms are described in turn. In each case, key information can be derived from the magnetic circular dichroism (MCD) spectrum which cannot easily be obtained from the electronic absorption spectrum alone. The use of MCD spectroscopy is not limited to a quantitative analysis of the three Faraday terms, however. Chapter 10 will describe how MCD spectroscopy and molecular orbital theory can be used to validate theoretical descriptions of the electronic structure of heteroaromatic and antiaromatic π-systems and to elucidate and rationalise key trends in their properties.

9.1 Examples of the Analysis of Faraday $\mathcal{A}_1$ Terms

Derivative-shaped Faraday $\mathcal{A}_1$ terms tend to dominate the MCD spectra when the excited states are orbitally degenerate. The intensity crossover point corresponds to the band centre of the absorption band. In the early years of MCD spectroscopy the technique was often used to derive values for the excited state magnetic moments of small high symmetry compounds and ion species by calculating the $\mathcal{A}_1/\mathcal{D}_0$ ratios, see equation (1.45), since this can be used to validate the accuracy of theoretical descriptions of the electronic structure. More recently, the technique has tended to be used to identify electronic origins in spectral envelopes of extensively overlapping bands on the basis of spectral band deconvolution analyses, because the derivative-shaped $\mathcal{A}_1$ terms associated with electronic transitions to orbitally degenerate excited states can be readily distinguished from the Gaussian-shaped Faraday $\mathcal{B}_0$, which are often observed for vibrational bands.

Circular Dichroism and Magnetic Circular Dichroism Spectroscopy for Organic Chemists
By Nagao Kobayashi, Atsuya Muranaka and John Mack

Published by the Royal Society of Chemistry, www.rsc.org

9.1.1 Cyclononatetraenide Anion[1]

The anion of the nine-membered cyclic π-system, $C_9H_9^-$, has been shown to have a planar nonagonal structure based on nuclear magnetic resonance (NMR) studies.[2] The electronic absorption spectrum of its alkali metal salts in tetrahydrofuran contains only two structureless absorption bands, at 250 nm (ε = 100 000) and 320 nm (ε = 8900). The bands at shorter and longer wavelengths were assigned, respectively, as $^1A_1' \rightarrow {}^1E_1'$ and $^1A_1' \rightarrow {}^1E_4'$ transitions by molecular orbital calculations.[3] Since the groundstate is nondegenerate and the excited states are doubly degenerate, Faraday $\mathcal{A}_1$ terms are observed in the MCD spectrum, Figure 9.1 with inflection points at 250 and 320 nm, which

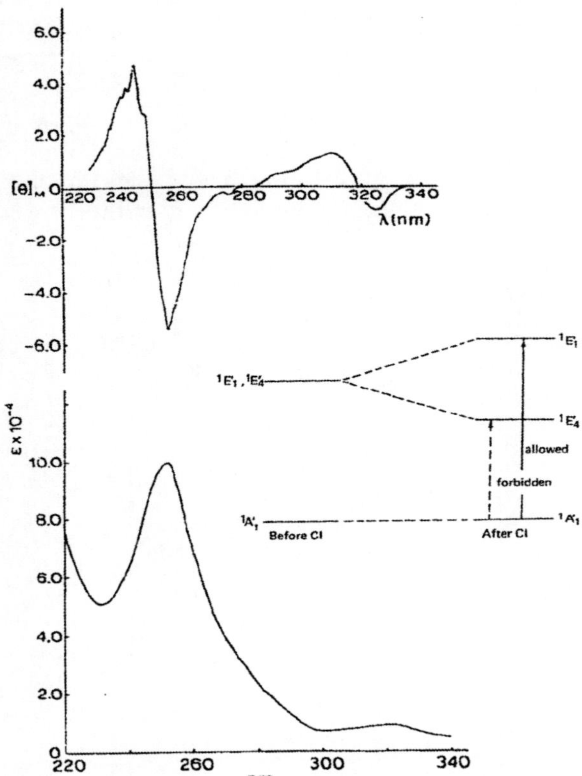

Figure 9.1 MCD (top) and absorption (bottom) spectra of the cyclononatetraenide anion at room temperature.[1] Molecular orbital calculations for the cyclononatetraenide anion assuming a nine-fold rotation axis assign the weaker band at longer wavelength corresponding to the first positive Faraday $\mathcal{A}_1$ term to a forbidden $^1A'_1 \rightarrow {}^1E'_4$ transition and the more intense band in the shorter wavelength region corresponding to the first positive Faraday $\mathcal{A}_1$ term to a forbidden $^1A'_1 \rightarrow {}^1E'_1$ transition. [Reprinted with permission from *Chem. Phys. Lett.*, 1974, **25**, 546. Copyright 1993, Elsevier.]

coincide with the absorption maxima of the $^1A_1' \rightarrow {}^1E_1'$ and $^1A_1' \rightarrow {}^1E_4'$ transitions. The magnitude of the magnetic moments of the excited states was evaluated by extracting the experimental $\mathcal{A}_1/\mathcal{D}_0$ values using Briat's method, described in Section 1.3.4.

9.1.2 Cycloheptatrienyl (Tropylium) Cation[4]

The tropylium cation, $C_7H_7^+$, is one of the most interesting monomeric odd-membered non-benzenoid aromatic hydrocarbons. The π-system is strongly aromatic due to the delocalisation of the six π-electrons. Molecular orbital calculations at various levels of theory predict the presence of an allowed transition at *ca.* 5.8–6.4 eV ($^1A_1' \rightarrow {}^1E_1'$) and a forbidden transition at *ca.* 4.6 eV ($^1A_1' \rightarrow {}^1E_3'$). Two bands are observed in the optical spectrum at 217 nm ($\varepsilon = 41\,000$) and 274 nm ($\varepsilon = 4350$), which were assigned, respectively, to the allowed and forbidden transitions on the basis of their relative intensities. If these assignments are correct, a derivative-shaped Faraday $\mathcal{A}_1$ term should be observed with inflection points corresponding to the absorption maxima. A negative $\mathcal{A}_1$ term is observed at 274 nm, Figure 9.2 with an $\mathcal{A}_1/\mathcal{D}_0$ value of $-2.80\ \beta$, while the $[\mathcal{B}_0 + C_0/kT]/\mathcal{D}_0$ ratio was found to be negligibly small.

9.1.3 C_{60} Fullerene[5,6]

Figure 9.3 contains the absorption and MCD spectra of C_{60} isolated in Ar matrices. The absorption bands at 326, 253 and 208 nm are very intense.

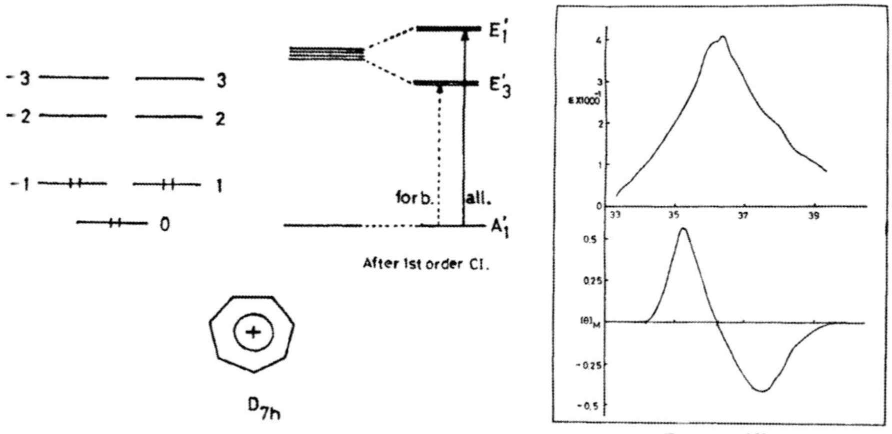

Figure 9.2 The π-MOs labelled by the M_L quantum number and the electronic states of the tropylium ion (left). MCD (bottom right) and absorption (top right) spectra of the tropylium cation in 0.1 N HCl.[4] [Reprinted with permission from *Bull. Chem. Soc. Jpn.*, 1972, **45**, 962. Copyright 1972, The Chemical Society of Japan.]

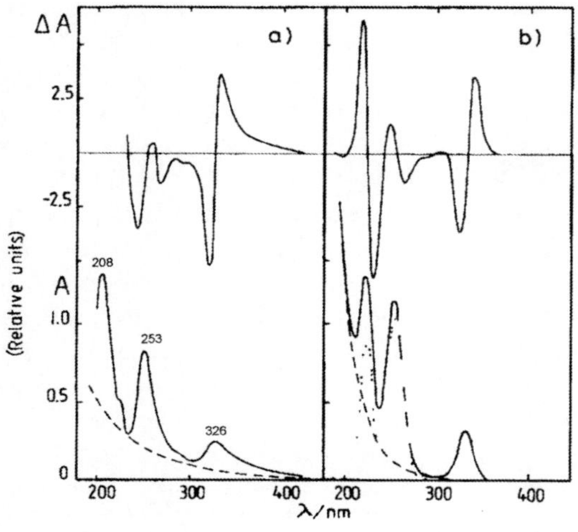

Figure 9.3 The experimental (a) and the theoretical (b) MCD and absorption spectra of C_{60}. Solid and dotted lines in the theoretical absorption spectrum denote the presence and absence of a background addition.[6] [Reprinted with permission from *Chem. Phys.*, 1993, **172**, 277. Copyright 1993, Elsevier.]

The 253 nm band has a molar extinction coefficient of ~175 000 M^{-1} cm^{-1} in hexane.[7] Given that C_{60} belongs to the I_h point group, and only $^1A_g \rightarrow {}^1T_{1u}$ transitions are electric dipole allowed in that context, the assignment of the three intense bands is clear-cut, with Faraday $\mathcal{A}_1$ MCD terms anticipated in each case. $\mathcal{A}_1$ terms are observed at 326 and 253 nm, but the band at 208 nm could not be recorded owing to the limits of the available instrumentation. The sign and magnitude of the $\mathcal{A}_1/\mathcal{D}_0$ ratio for these transitions were found to be consistent with values calculated using the self-consistent field Pariser-Parr-Pople configuration interaction (SCF-PPP-CI) method, Figure 9.3. Although there is significant Faraday $\mathcal{B}_0$ term intensity in the MCD spectrum, the detection of the Faraday $\mathcal{A}_1$ terms was relatively straightforward.

9.1.4 Identification of the Electronic Origin of the S1 State of Zinc Tetraphenyltetraacenaphthoporphyrin[11]

Method of moments analysis was the first approach used to derive $\mathcal{A}_1/\mathcal{D}_0$, $\mathcal{B}_0/\mathcal{D}_0$ and $\mathcal{C}_0/\mathcal{D}_0$ ratio values to produce accurate quantitative values for ground and excited state magnetic moments. Unfortunately, the technique is only suitable for baseline-to-baseline spectral band envelopes. Spectral deconvolution with Gaussian-shaped curves tends to be used to extract $\mathcal{A}_1/\mathcal{D}_0$, $\mathcal{B}_0/\mathcal{D}_0$ and $\mathcal{C}_0/\mathcal{D}_0$ ratios from spectral data sets that lack discrete baseline-to-baseline band envelopes.[8] For example, in the 1980s, Stillman and coworkers developed specialist in-house software to obtain the band energies and polarisations of the major spectral bands within the highly overlapping band envelopes that are often encountered in porphyrinoid UV–visible absorption and MCD spectral data.[9] Band fitting can also now be carried out using commercial software.[10] Since the MCD and UV–visible absorption spectra arise from the same set of spectral bands, the measurement of a second data set removes much of the ambiguity normally associated with band deconvolution techniques. Under the rigid-shift approximation it can be assumed that the application of the applied magnetic field makes no difference to the spectral bandshape function, so the same set of bandwidth and band centre parameters can be used to fit both spectra. Identifying the main electronically allowed bands can be challenging when spectral bands overlap extensively. This problem can be resolved by making use of the distinctive derivative shape of the Faraday $\mathcal{A}_1$ term. For example, in the visible region of the spectrum of zinc tetraphenyltetraacenaphthoporphyrin (ZnTPTANP), Figure 9.4, MCD spectroscopy

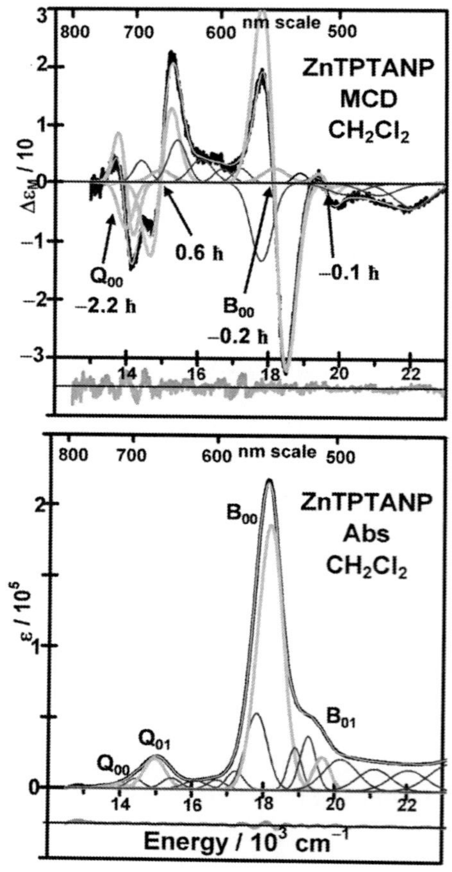

Figure 9.4 Spectral deconvolution analyses of the absorption and MCD spectra of zinc tetraphenyltetraacenaphthoporphyrin recorded in CH_2Cl_2.[11]

provides definitive evidence that the electronic origin of the band associated with the S1 excited state lies at 717 nm. In the context of the electronic absorption spectrum, this lies in what appears to be a set of hot bands associated with the more intense vibrational band that corresponds in energy terms to a second higher energy Faraday $\mathcal{A}_1$ term at *ca.* 660 nm.[11]

9.2 Examples of the Analysis of Faraday $\mathcal{B}_0$ Terms

Faraday $\mathcal{B}_0$ terms are always present in MCD spectra due to the field-induced mixing of close-lying states linked by mdtms. In the context of molecules and complexes that lack a three-fold or higher axis of symmetry, there is a complete lifting of ground and excited state degeneracies and only Gaussian-shaped Faraday $\mathcal{B}_0$ terms are present in the MCD spectrum. When a low symmetry molecule is formed from a higher symmetry parent compound with orbitally degenerate $\pi\pi^*$ excited states, the derivative-shaped Faraday $\mathcal{A}_1$ terms are usually replaced by relatively intense pairs of coupled and oppositely signed Faraday $\mathcal{B}_0$ terms, see Figure 1.23. In the next chapter a theoretical framework known as Michl's perimeter model will be described, which can be used to derive information from sign sequences observed in the Faraday $\mathcal{A}_1$ and $\mathcal{B}_0$ terms of aromatic and heteroaromatic π-systems. On a more basic level the presence of coupled pairs of oppositely signed Faraday $\mathcal{B}_0$ terms in an MCD spectrum can be used to identify the main electronic bands in the UV-visible absorption spectrum. The absence of a $1/kT$ temperature dependence in the spectra of high symmetry complexes can also be used to confirm the presence of an orbitally nondegenerate groundstate.

9.2.1 Analysis of the Symmetry-split Excited States of Azaporphyrins[12]

The metal complexes of radially symmetrical porphyrins, tetraazaporphyrins and phthalocyanines have approximate D_{4h} symmetry based on the presence of a central C_4 axis. This leads to an orbitally nondegenerate $^1A_{1g}$ groundstate and doubly degenerate ^1E_u $\pi\pi^*$ excited states, which are optically accessible on the basis of electronically allowed x/y-polarised transitions. The MCD spectra of porphyrinoids are normally dominated by two Faraday $\mathcal{A}_1$ terms associated with a weaker Q band at longer wavelengths and a more intense B band in the UV region. When the molecular symmetry is lowered through the fusion of benzene rings to the ligand periphery, the Q band splits into two x- and y-polarised components, and coupled oppositely signed Faraday $\mathcal{B}_0$ terms are observed in the MCD spectrum. Figure 9.5 contains the structures of a series of low symmetry tetraazaporphyrin compounds, which have been studied by MCD spectroscopy, Figure 9.6.[12] The MCD spectra of the D_{4h} symmetry **Zn** and **4Zn** complexes are dominated by Faraday $\mathcal{A}_1$ terms, since the main $\pi\pi^*$ excited states are doubly degenerate. There is no obvious splitting of the Q band in the **2AdZn** spectrum. Simultaneous spectral band deconvolution analysis of the absorption and MCD spectra (dotted lines) indicates that the

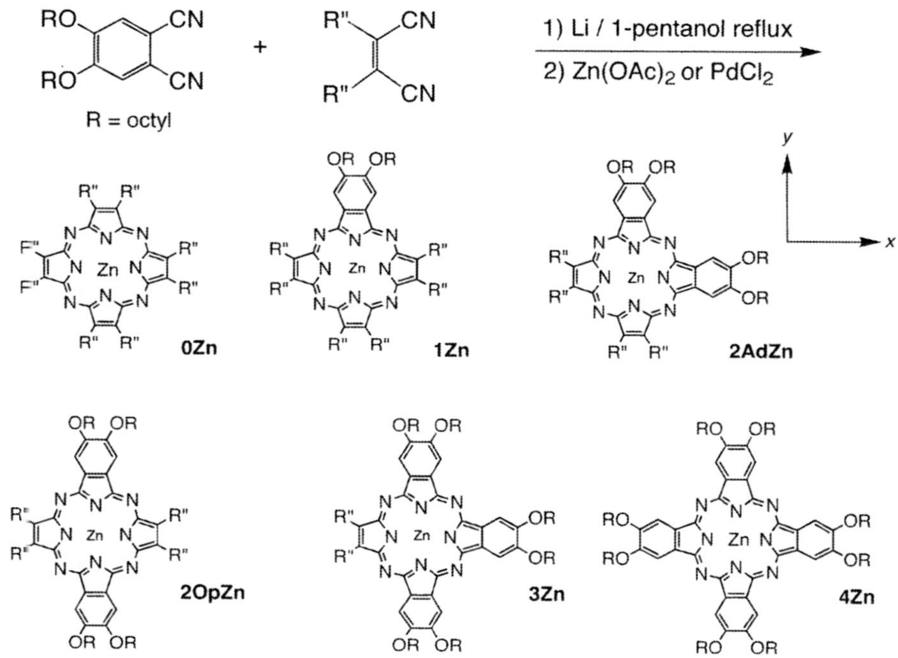

Figure 9.5 Molecular structures of low-symmetry tetraazaporphyrin (TAP) derivatrives, together with their abbreviations. The numbers in the compound names indicate the number of fused benzene rings, while the Zn implies that the central metal ion is Zn(II).[12]

intense absorption band at 654 nm is formed by the superimposition of two close-lying bands. The Q band region of the MCD spectrum appears to contain a derivative-shaped Faraday $\mathcal{A}_1$ term, despite the fact that the absence of a three-fold or higher symmetry axis should result in a complete lifting of ground and excited state degeneracies. Two close-lying Faraday $\mathcal{B}_0$ terms of opposite sign are observed in the MCD spectrum because there is a very small zero field splitting of the Q_x and Q_y excited states. This type of signal is sometimes referred to as a pseudo-Faraday $\mathcal{A}_1$ term. As would be predicted on the basis of first order perturbation theory,[13] the values for splitting of the Q_x and Q_y bands of **1Zn** and **3Zn** are of similar magnitude. An important feature of the spectrum of **3Zn** is that the position of the split components can be definitively identified on the basis of the MCD data. The absorption band centre at 690 nm is readily identified as the lower energy Q(0,0) band. In the absence of MCD data there would be no way to determine whether the band centre of the second component lies at 628 or 642 nm. Since the intensity mechanism of Faraday $\mathcal{B}_0$ terms, equation (1.38), is based on the field-induced mixing of states linked by a mdtm,[14,15] the −, + sign sequence observed in the Faraday $\mathcal{B}_0$ terms at 689 and 643 nm, analogous to what is observed in the Faraday $\mathcal{A}_1$ terms of **Zn** and **4Zn**, provides direct spectroscopic evidence that the band centre of the higher

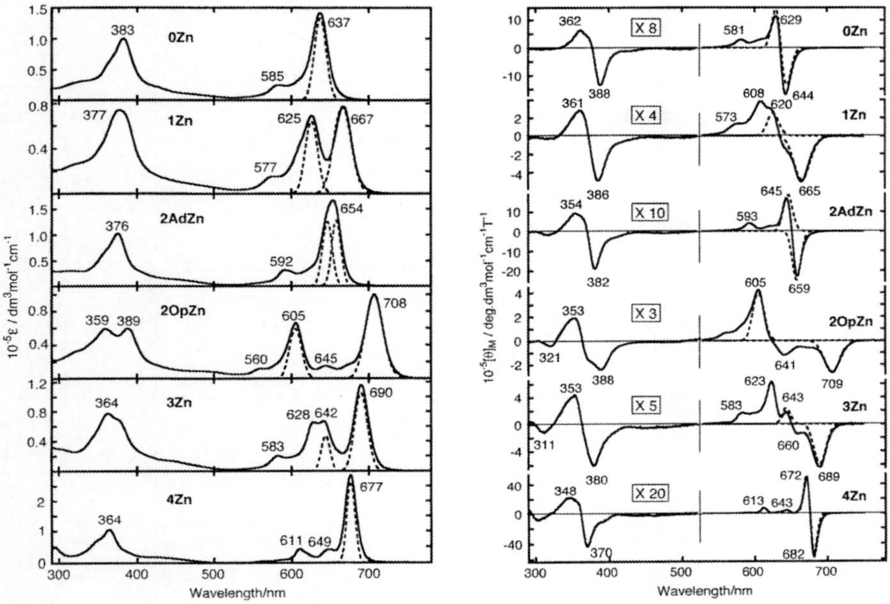

Figure 9.6 Electronic absorption (left) and MCD spectra (right) of the zinc complexes shown in Figure 9.5.[12]

energy component lies at 642 nm. The largest splitting is observed for **2OpZn**.[13] This leads to a drop in intensity, as a result of a large energy split between the states that are mixed in the applied magnetic field (ΔE_{KJ}), see equation (1.38). The trends observed in the calculated spectra, such as the splitting of the Q bands and the relative intensities of the Q_x and Q_y bands, Figure 9.7 are consistent with those observed in the absorption spectra of these complexes, Figure 9.6.

9.2.2 Analysis of Data Recorded at Cryogenic Temperature: Phthalocyanine Anion Radicals[17]

When the MCD spectrum is dominated by Gaussian-shaped bands at room temperature with band centres aligned with those in the electronic absorption spectra, temperature dependence studies can be used to determine whether Faraday $\mathcal{B}_0$ or $\mathcal{C}_0$ terms are present. The MCD spectroscopy of porphyrinoid anion radicals provides a typical example. On the basis of group theory considerations alone, the addition of an electron to the $1e_g$* LUMO of a phthalocyanine π-system should result in an orbitally degenerate ^2E_g groundstate, and attempts were made to interpret the UV-visible absorption data of metal phthalocyanine anion radicals ($[MPc(-3)]^{\bullet -}$) on this basis in the absence of MCD spectral data.[16] No $1/kT$ temperature dependence was observed,

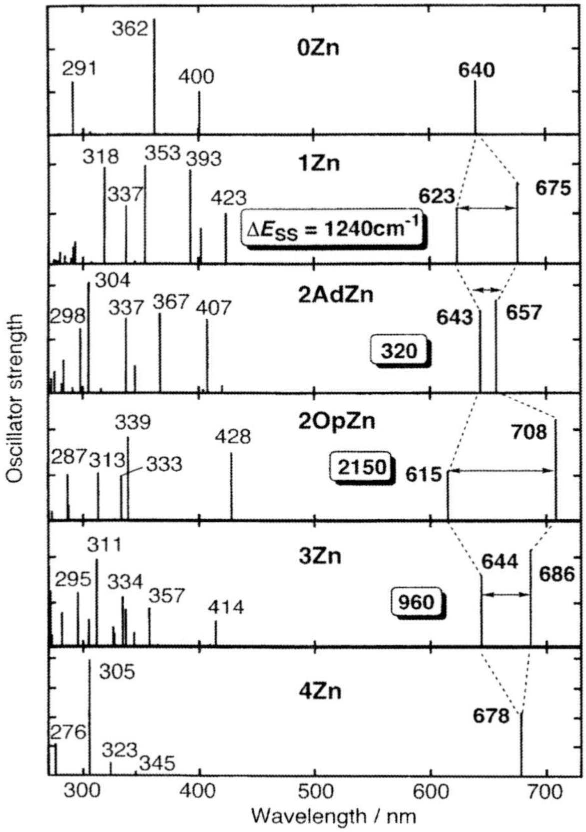

Figure 9.7 Absorption wavelength and oscillator strengths obtained by configuration interaction (CI) calculations on the complexes in Figure 9.5.[12]

Figure 9.8, when room temperature and low temperature MCD spectra were recorded subsequently for $[MgPc(-3)]^{\bullet -}$, however.[17] This provided definitive and unambiguous spectroscopic evidence that the orbitally degenerate groundstates of phthalocyanine anion radicals are subject to a static Jahn-Teller distortion effect at zero-field, which lifts the orbital degeneracy, because Faraday C_0 terms would be observed if orbital degeneracy was retained. The spectrum is, instead, dominated by coupled oppositely signed Faraday $\mathcal{B}_0$ terms, as is the case with low symmetry azaporhyrins.

A complete spectral band deconvolution analysis was carried out for the electronic absorption and MCD spectra of the ligand-reduced $[ZnPc(-3)]^{\bullet -}$ species, Figure 9.9, recorded at cryogenic temperatures in vitreous samples formed by quenching dimethyl formamide and hydrazine hydrate solvent mixtures in liquid nitrogen. This eliminated the presence of hot bands associated with the partial population of higher energy vibrational levels of the

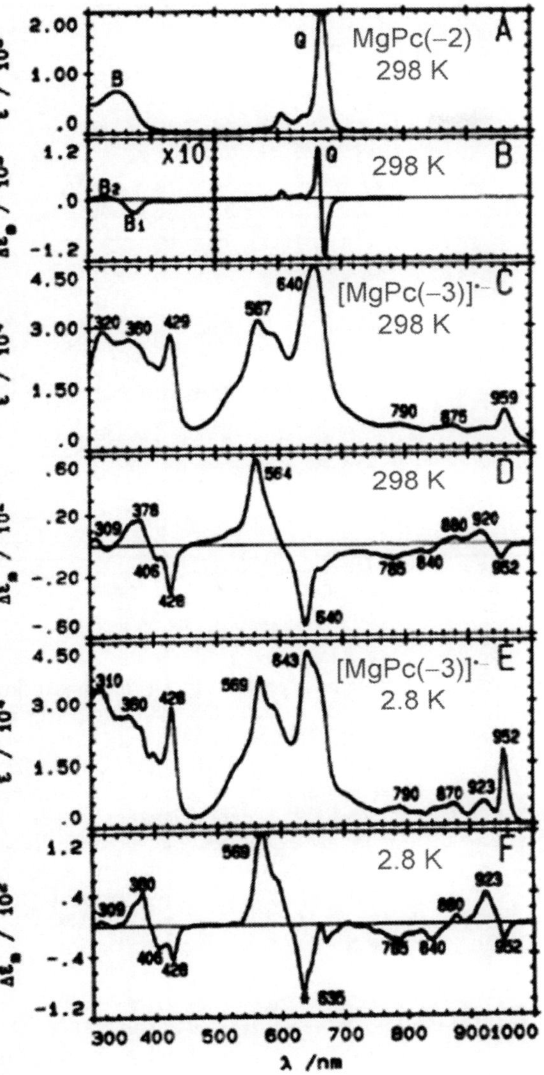

Figure 9.8 Absorption (A/C/E) and MCD (B/D/F) spectra of neutral MgPc(−2) (A/B) and one-electron-reduced $[MgPc(-3)]^{\bullet -}$ at room (C/D) and low temperature (E/F): (A/B) MgPc in DMF; (C) absorption and (D) MCD spectra of $[MgPc(-3)]^{\bullet -}$ recorded at 298 K by reduction of a solution of MgPc/DMF/0.1 M TPAP; (E/F) absorption spectrum of $[MgPc(-3)]^{\bullet -}$ recorded at 77 K and an MCD recorded at 2.8 K on a vitreous sample formed by reduction of MgPc/DMF-DMA/0.1 M TPAP.[17a] The shoulder at 645 nm on the negative MCD band is an artefact of the measurement and contributes to a low value for the peak maximum, indicated by *635. No $1/kT$ temperature dependence is observed in the MCD spectra so the MCD spectrum is clearly comprised of Faraday $\mathcal{B}_0$ terms. [Reprinted with permission from *Inorg. Chem.*, 1992, **31**, 1717. Copyright 1992, American Chemical Society.]

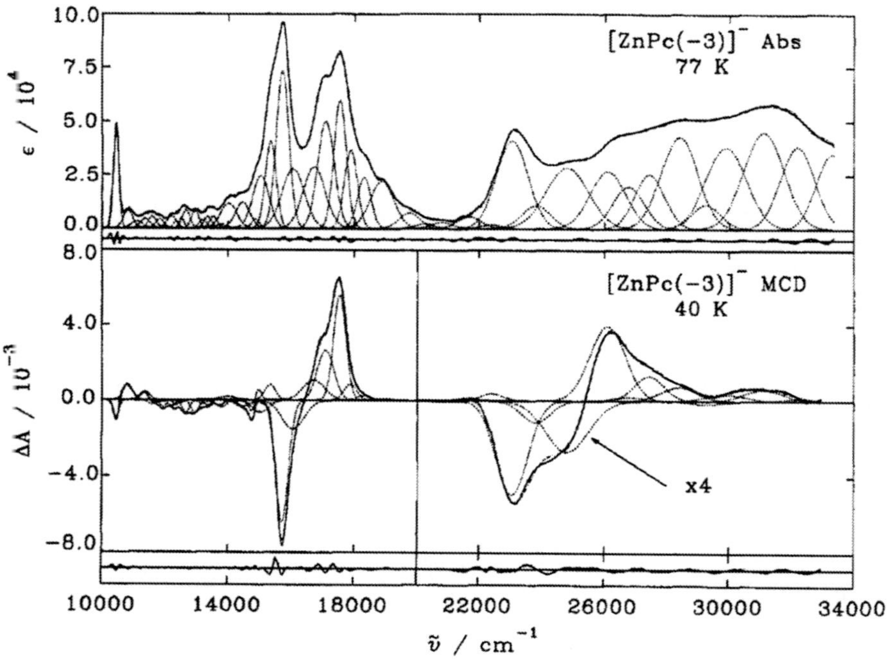

Figure 9.9 Complete set of 45 bands required to fill the absorption and MCD band envelopes of $[ZnPc(-3)]^{\bullet-}$. One additional band was required to account for the very intense Q(0,0) band of ZnPc(−2). Only a very minor impurity of ZnPc(−2) is present (*ca.* 1%). The absorption spectrum was recorded at 77 K, and the MCD spectrum at 40 K. The bands were fit using Gaussian-shaped components with the identical band centres and bandwidths for pairs of absorption and MCD bands. The bandwidths for the three regions (960, 640 and 400 nm) are clearly different. The weak, positive $\mathcal{A}_1$ term located at 14 860 cm^{-1} is due to residual neutral ZnPc(−2). [Reprinted with permission from *J. Am. Chem. Soc.*, 1994, **116**, 1292. Copyright 1994, American Chemical Society.]

groundstate, which often have significant intensity in spectra measured at room temperature in solution. When the bandwidths are plotted against the energy of the band centres, Figure 9.10, marked differences are observed in sets of electronic and vibrational bands associated with different transitions, due to the effect of Heisenberg's uncertainty principle. Bandwidths increase markedly at shorter wavelengths as the lifetimes of the higher energy excited states become progressively shorter. It is often unwise, therefore, to use a fixed bandwidth to model optical spectra based on the results of a theoretical calculation. Very large bandwidths of *ca.* 3000 cm^{-1} are often the default setting in commercially available software packages, which are based on the values that would be most suitable to simulate the spectra of small molecules which absorb primarily in the UV region. Much smaller bandwidths are normally required at longer wavelengths, because the excited states associated with the transitions have

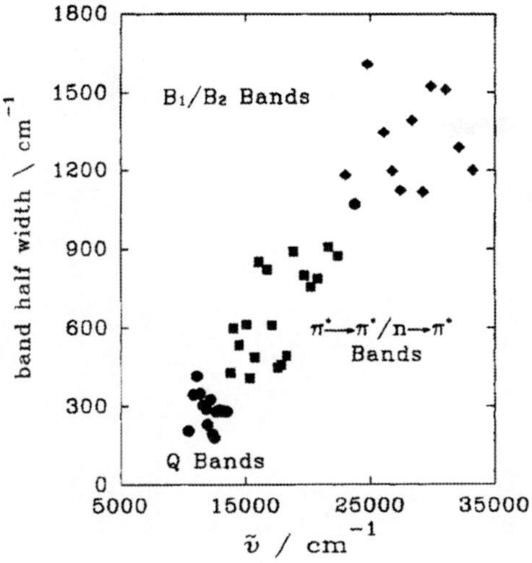

Figure 9.10 Scatter plot of the bandwidths used in the absorption band fit shown in Figure 9.9 as a function of the energy of the band centre. The symbols indicate bandwidths taken from the following regions of the spectrum: (circles) Q band near 960 nm; (squares) $\pi^* \rightarrow \pi^*$ and $n \rightarrow \pi^*$ bands near 640 nm; (diamonds) B1/B2 bands near 380 nm. [Reprinted with permission from *J. Am. Chem. Soc.*, 1994, **116**, 1292. Copyright 1994, American Chemical Society.]

significantly longer lifetimes. For example, Figure 9.10 suggests that bandwidths of 100–150 cm^{-1} at *ca.* 1000 nm, 400–450 cm^{-1} at *ca.* 700 nm, 700–900 cm^{-1} at *ca.* 500 nm, 1000–1200 cm^{-1} at *ca.* 400 nm and 1500–1700 cm^{-1} at *ca.* 300 nm would be appropriate for the spectra of $[MPc(-3)]^{\bullet-}$ species recorded at cryogenic temperatures. It should be noted that significantly larger bandwidth values would be required to fit the room temperature data of these species owing to the presence of hot bands and solvation effects.

9.3 Examples of Analysis of Faraday C_0 MCD Terms

9.3.1 The "Fingerprint" Approach

Since the mid-1960s, MCD spectroscopy has often been used to study haem proteins, such as myoglobin and hemoglobin and other biologically significant transition metal complexes. MCD spectroscopy has consistently provided key ground and excited state degeneracy information, which cannot easily be derived from an analysis of the UV-visible absorption spectra and theoretical calculations. Major changes are observed in the MCD spectra as a result of the differing band morphologies of the three Faraday terms, while in contrast only relatively minor changes are typically observed in the band centres and intensities of the UV–visible absorption spectra. Since there is a partially filled set of

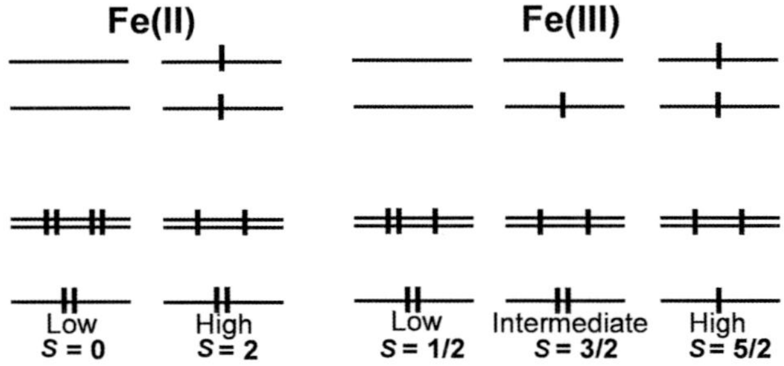

Figure 9.11 The low, intermediate and high spin states for Fe(II) and Fe(III) based on a tetragonal environment.

d orbitals, Figure 9.11, ligand to metal charge transfer (LMCT) and metal to ligand charge transfer (MLCT) bands are observed in addition to the main $\pi \rightarrow \pi^*$ bands. The redox chemistry of transition metal porphyrinoid complexes is significantly more complex than that of main group metal and closed shell d^{10} metal complexes. The spin state of Fe(II) and Fe(III) porphyrin complexes can be low, intermediate or high, Figure 9.11, depending on the nature of the axial ligation along the *z*-axis. The configuration interaction between the main $\pi\pi^*$ and CT states results in so called "fingerprint" patterns,[18] which can be used to derive information about several key points: (i) the nature and number of ligands attached to the iron; (ii) the spin and oxidation states of the iron; (iii) the ease with which the iron can be oxidised or reduced; (iv) whether axial ligands can approach and bind the iron. A qualitative analysis of the distinctive MCD band morphologies in room temperature spectral data can provide this information based on earlier measurements on model complexes, even in the absence of quantitative studies of the temperature dependence of Faraday C_0 term signals which may be present.

Recently this approach has been used to explore the haem binding properties of iron-regulated surface determinant proteins in the human pathogen *Staphylococcus aureus*.[19] Despite the fact that the spectrum is clearly consistent with the presence of a high spin d^5 Fe(III) ion on the basis of the "fingerprint" approach, adding excess CN^-, which would normally lead to a low spin spectrum, makes no discernible difference, Figure 9.12. It was only possible to ligate a CO ligand after reduction to Fe(II) with sodium dithionite. This demonstrated that haem is tightly bound and not readily accessible to small inorganic anions. The Fe environment was identified identified as being 5-coordinate, with tyrosine as the probable axial ligand, on the basis of the wavelengths and the morphologies of the main spectral bands. When the X-ray structure was subsequently reported,[20] the crystal structure of the NEAT domain–haem complex of Isd A revealed that the iron is indeed 5-coordinate, with the phenolic oxygen of a tyrosine residue acting as an axial ligand in a binding pocket containing a series of hydrophobic residues. This illustrated the value of the "fingerprint approach" for analysing MCD spectral data when no X-ray structure is available.

9.3.2 Analysis of Data Recorded at Cryogenic Temperatures

The absorption bands that lie between 1000 and 2000 nm in the near IR region of low spin d^5 Fe(III) haem spectra have been assigned to an LMCT transition. Under D_{4h} symmetry, the a_{1u} HOMO of the porphyrin ligand in main group metal complexes is occupied by two electrons, so the groundstate is non-degenerate and no Faraday C_0 terms are anticipated. In contrast, the low spin Fe(III) ion of iron porphyrin complexes is known to lie at the centre of the central cavity so the MOs of the ligand can readily interact with the *d* orbitals of the central metal. The groundstate can therefore be expressed as the product of the wavefunctions of the metal and the ligand. The groundstate is doubly degenerate, based on whether the d_{xz} or d_{yz} orbital is singly occupied. Since the lowest energy LMCT transition results in an $(a_{1u})^1(d_{xz})^2(d_{yz})^2$ electronic configuration, the associated excited state is nondegenerate. A Faraday C_0 term is, therefore, anticipated with a band morphology that is similar to that of the electronic absorption band as is indeed observed for Fe(III) cytochrome *c* at alkaline pH using a stream of cold nitrogen gas, Figure 9.13.[21] As the inset demonstrates, an unambiguous $1/kT$ dependence is observed in the MCD intensity. It should be noted that the use of expensive optical cryostats was not required in this context to analyse the MCD spectra of these proteins successfully.

The MCD data of d^5 Fe(III) haem in the near IR region, Figure 9.14, have proven to be the ideal probe for determining the extent of the rhombic splitting of the d_{yz} and d_{xz} orbitals, Figure 9.15. When electron paramagnetic resonance (EPR) spectroscopy is used, the analysis of the data can be problematic when the g_z component of the g tensor is greater than 3.4, because the other two components g_x and g_y, become difficult to detect experimentally. The detection of the absorption band associated with the LMCT transition becomes increasingly difficult experimentally as its wavelength lies towards 2000 nm, and this spectral region is partially obscured by the presence of vibrational overtone peaks arising from the C–H, O–H and N–H bonds of the protein. Although the exchange of protons for deuterons would cause the vibrational peaks to move to longer wavelengths, protons in proteins are not always easily exchanged. The problem can be solved by measuring MCD spectra because the intensity depends in part upon the magnetic moments of the particle involved in the transition. The MCD signals of vibrational transitions are several orders of magnitude weaker than those of electronic transitions, since the magnetic moments of nuclei are several orders of magnitude smaller than those of an electron. By making measurements at cryogenic temperatures the MCD intensity of the LMCT band can be further enhanced relative to the vibrational bands owing to the $1/kT$ temperature dependence of C_0 terms. A strong correlation was obtained between the energy of the positive hole on the Fe(III) ion, as determined experimentally from EPR g factor values, and the energy of

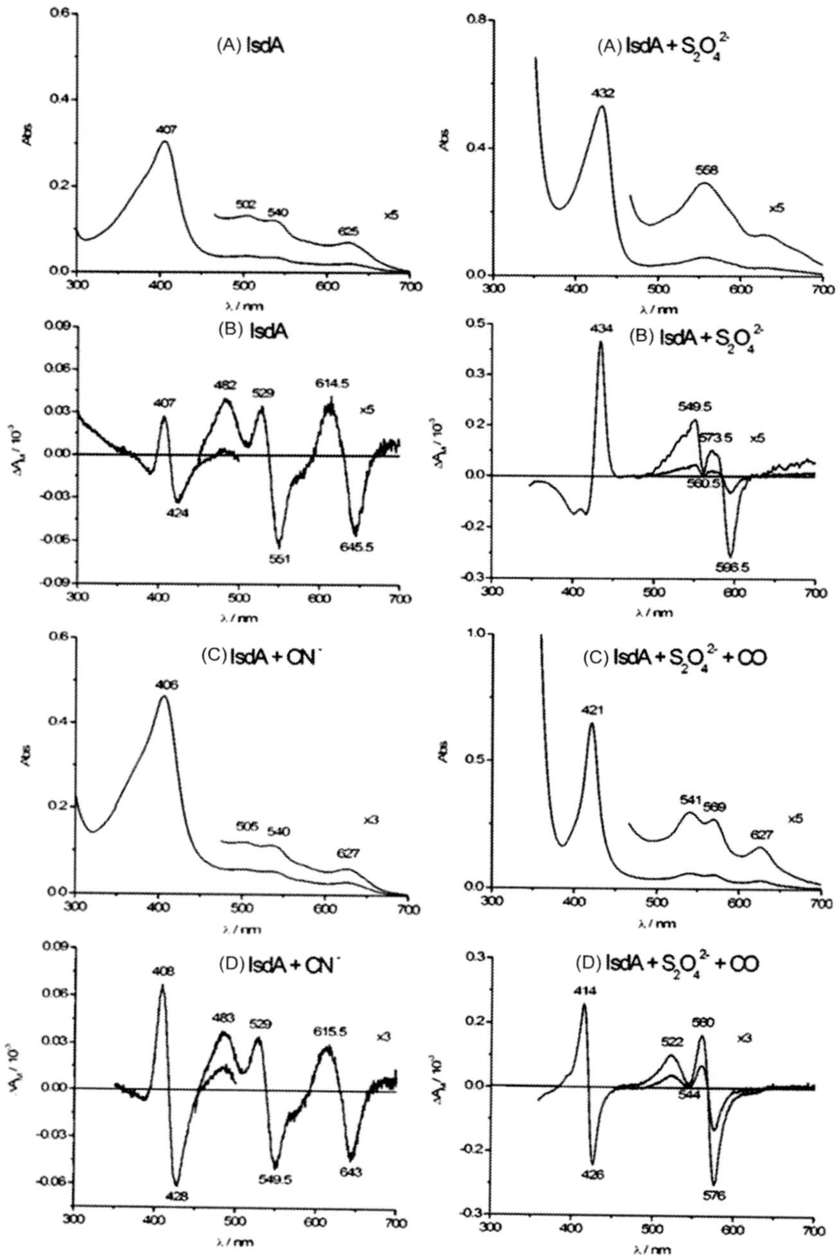
(A) IsdA
407
502
540
625
x5
Abs
λ / nm
(A) IsdA + $S_2O_4^{2-}$
432
558
x5
(B) IsdA
407
424
482
529
551
614.5
645.5
x5
ΔA_M / 10^3
(B) IsdA + $S_2O_4^{2-}$
434
549.5
573.5
560.5
596.5
x5
(C) IsdA + CN^-
406
505
540
627
x3
(C) IsdA + $S_2O_4^{2-}$ + CO
421
541
569
627
x5
(D) IsdA + CN^-
408
428
483
529
549.5
615.5
643
x3
(D) IsdA + $S_2O_4^{2-}$ + CO
414
426
522
544
560
576
x3

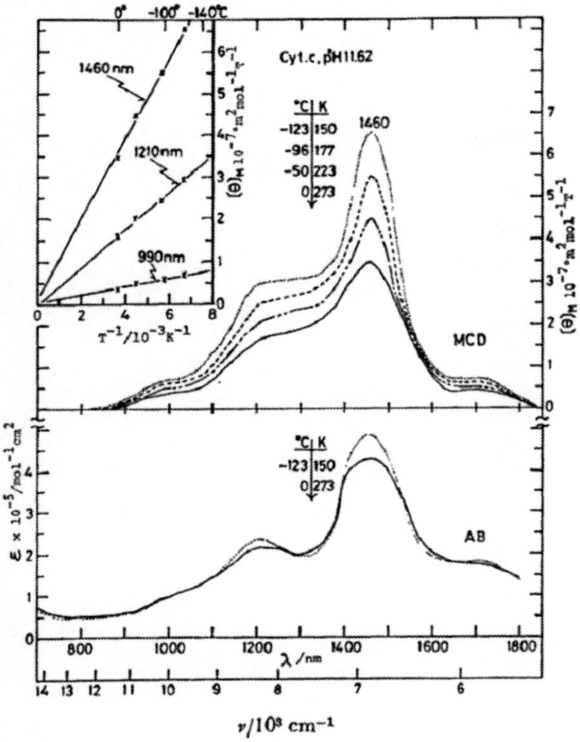

Figure 9.13 MCD and absorption spectra of Fe(III) cytochrome *c* at pH = 11.62. A cell length of 3 mm was used along with a concentration of 4.5 × 10^{-3} M and a field strength of 1.47 T. The inset shows the MCD magnitudes of the peaks at 1460, 1210 and 990 nm *vs.* the reciprocal of the absolute temperature. Reproduced with permission from reference 22a.

Figure 9.12 Absorption and MCD spectra of ferric IsdA (A and B) as isolated at pH = 7.0 and (C and D) following the addition of crystalline sodium cyanide in small aliquots up to an extremely large excess. No changes were observed in the spectra over a period of hours (left). (A and B) Absorption and MCD spectra of ferrous IsdA with the addition of sodium hydrosulfite (right). Crystalline hydrosulfite was added in small aliquots as a reductant until there were no further changes in the absorption and MCD spectra. The band at 431 nm was monitored and maximised. (C and D) Absorption and MCD spectra of ferrous (d^6) IsdA with the addition of sodium hydrosulfite and carbon monoxide. CO was bubbled slowly into a sealed solution of reduced IsdA until no further changes in the intensity of the B band at 421 nm were observed.[19] [Reprinted with permission from *Biochemistry*, 2006, **45**, 12867. Copyright 2006, American Chemical Society.]

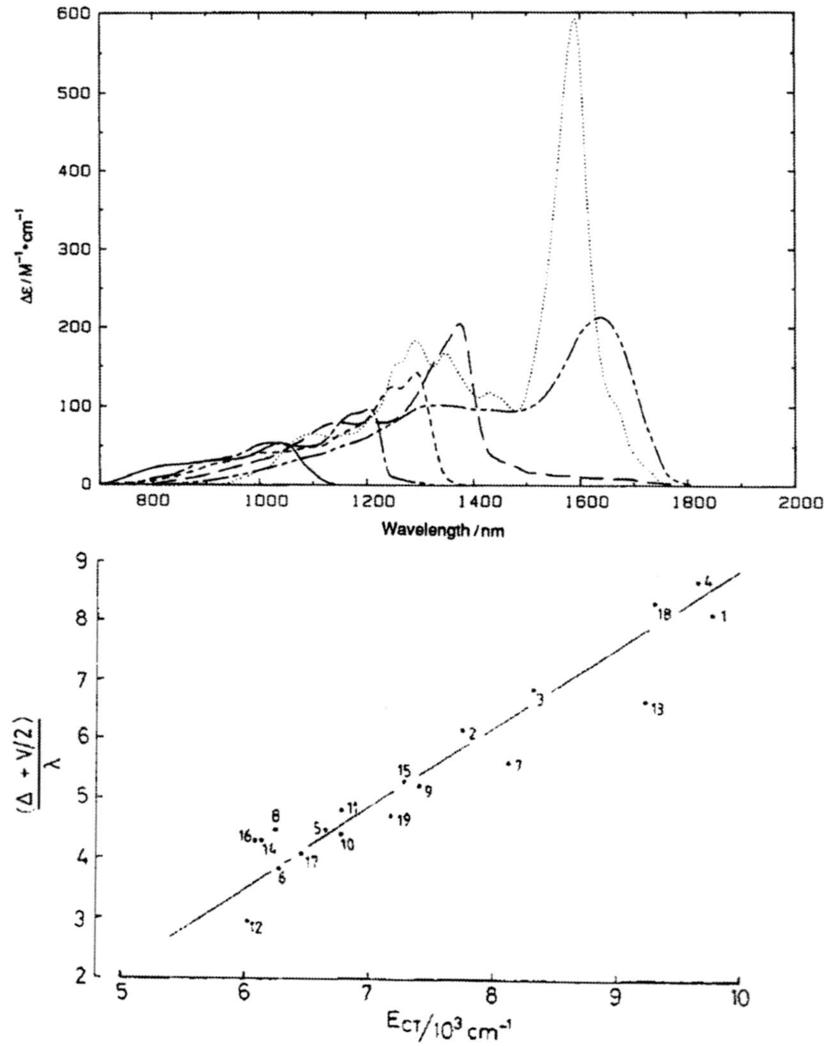

Figure 9.14 Near-infrared MCD spectra of ferric haem adducts of metmyoglobin and metleghemoglobin (soybean) (top).[21b] Spectra were recorded at 4.2 K or 5 T with band centres from left to right for metMb–SC_3H_7, metMb–SH, metMb–N_3, Lb–imidazolate, metMb–CN and Lb–imidazole. A plot of $(\Delta + V/2)/\lambda$, the energy separation between the d_{yz} and d_{xy}, orbitals (where Δ, V and λ are the axial and rhombic ligand field parameters and the spin-orbit coupling, respectively), versus the energy of the near-infrared region LMCT band (E_{CT}) (bottom). [Reprinted with permission from *J. Am. Chem. Soc.*, 1990, **112**, 5003. Copyright 1990, American Chemical Society.]

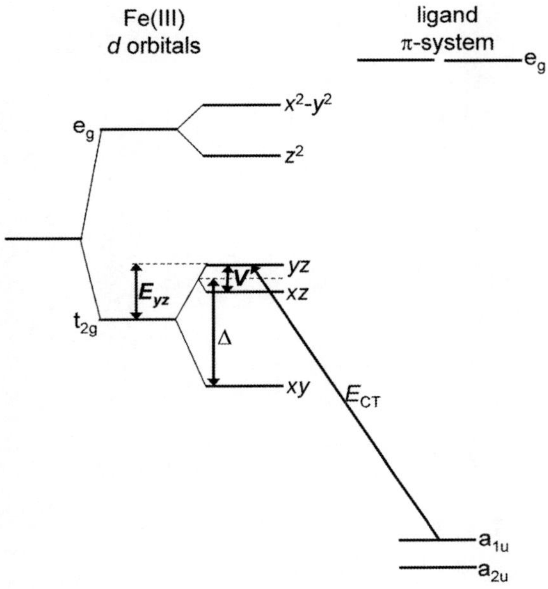

Figure 9.15 An energy level diagram for the *d* orbitals of the low spin ferric (d^5) ion and the π-MOs porphyrin ligand showing the lowest energy LMCT transition predicted between the porphyrin π-MOs and the *d* orbitals.[21b] The definitions of E_{CT}, E_{yz}, Δ and *V* are also indicated.

the CT band as identified by MCD spectroscopy. The numbered examples in Figure 9.14 all contain the protohaem IX ring, with histidine as the proximal ligand and a further ligand bound in the second axial position with a range of crystal field strengths covering much of the spectrochemical series. This correlation placed the assignment of the near$1/kT$ IR CT bands of low spin Fe(III) haemoproteins on a more secure footing and provides a useful linear relationship which can be used to identify the axial ligand.

9.3.3 The Quantitative Analysis of Magnetisation Curves

Magnetisation curves[23] can be used to derive quantitative information about the groundstate g factor values and spin states of paramagnetic metal sites in bioproteins on the basis of the theoretical expressions described in equations (1.51–1.55),[24] Figure 9.16. The key advantage of MCD spectroscopy over magnetic susceptibility measurements and Mossbauer and EPR spectroscopy in this regard is that individual metal centres can be studied when a specific transition absorbs at a discrete wavelength, while the other approaches provide bulk property values for the entire sample. Analysis of the $S=2$ groundstates associated with high spin d^6 configurations of Fe(II) are considerably more challenging than systems based on Kramers doublet, because multiple nested

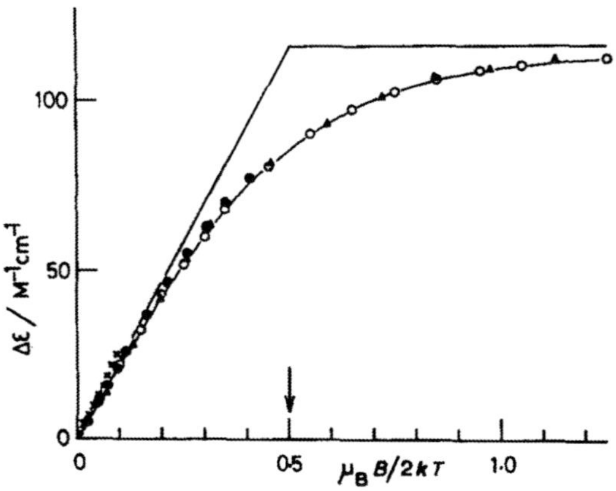

Figure 9.16 MCD magnetisation curves (*i.e.* plots of peak to trough intensity value for the B band in the MCD spectrum, $\Delta\varepsilon$ *vs* $\mu_B/2kT$) for reduced *Cl. pasteurianum* ferredoxin.[23b] $\lambda = 530$ nm. ▲, 1.52 K; ○, 4.22 K, and ×, 18.6 K. At each temperature the magnetic fields used were: 0.29, 0.59, 0.88, 1.16, 1.42, 2.06, 2.66, 3.24, 3.81, 4.39, 5.10 T, the theoretical curve (—o—) was computed with $g_{\parallel} = 2.06$, $g_{\perp} = 1.90$ based on equations (1.51–1.55). The arrow indicates the intercept value, I = 0.50, see Figure 9.29. [Reprinted with permission from *Biochim. Biophys. Acta*, 1991, **667**, 433. Copyright 1981, Elsevier.]

magnetisation curves are observed, depending on the temperature and field strength values. During the 1990s, variable temperature variable field (VTVF) techniques were developed for analysis of the MCD spectra of the non-Kramers doublet groundstates of non-haem iron enzymes.[25] The full details of these specialist analyses are beyond the scope of an introductory text on MCD spectroscopy, but the key advantage of this approach is that spin Hamiltonian parameters associated with a non-Kramers groundstate that is EPR silent can be derived from an analysis of excited state properties through an optical spectroscopy technique.

9.4 Analysis of MCD Spectra based on Molecular Orbital Calculations

One of the key uses of MCD spectroscopy in recent years has been to test the validity of density functional theory (DFT) based descriptions of the electronic structures of porphyrinoid complexes, because the additional state degeneracy and band polarisation information that can be derived from the three Faraday terms is unavailable from the UV-visible absorption spectrum alone. At present it is not possible to calculate values for the three Faraday terms using commercially available DFT based software packages. The first examples of MCD spectra calculated using TD-DFT have been reported recently by Ziegler and co-workers, who used the BP86 functional of the Amsterdam Density Functional suite.[26] At least in the context of small inorganic complexes the results appear to be quite promising.[27] In the case of magnesium and zinc porphyrin, however, a second intense electronic band has been predicted in the B band region due to the presence of a close-lying excited state associated with the $1b_{2u} \rightarrow 1e_g{}^*$ one-electron transition[28] where only one $\mathcal{A}_1$ term has typically been observed experimentally. The sign sequence predicted for the two $\mathcal{A}_1$ terms in the B band region is opposite from that observed experimentally in the case of MgP. Coupled oppositely signed $\mathcal{B}_0$ terms, due to the second close-lying excited state, had to be invoked to account for this, despite the fact that a single intense $\mathcal{A}_1$ term has been observed consistently in this spectral region for a wide range of different porphyrin structures. The selection of the exchange–correlation (x–c) functional may have been the key issue in this regard. It was demonstrated recently[29] that calculations carried out with commonly used x–c functionals such as BP86 and B3LYP underestimate the energy of states associated with the $1b_{2u} \rightarrow 1e_g{}^*$ one-electron transition of free base porphyrins and their fused-ring-expanded analogues and that more accurate predictions are provided by a Coulomb-attenuated x–c functional (CAM-B3LYP), which incorporates an increasing fraction of Hartree-Fock exchange as the interelectronic separation increases. Once the code for calculating the three Faraday terms is implemented into commercially available software, sign sequences observed in the MCD spectrum and $\mathcal{A}_1/\mathcal{D}_0$, $\mathcal{B}_0/\mathcal{D}_0$ and/or $\mathcal{C}_0/\mathcal{D}_0$ ratio values will provide an additional spectroscopic basis for assessing the validity of DFT and TD-DFT based descriptions of the electronic structures and optical spectra of aromatic compounds and inorganic complexes. In the context of aromatic π-systems, older theoretical approaches such as Michl's perimeter model[30] will continue to provide a useful conceptual framework, however, for identifying and analysing key trends observed in the experimental data and in calculated spectra.[31] Michl's perimeter model will be described in depth in the next chapter.

References

1. A. Tajiri and M. Hatano, *Chem. Phys. Lett.*, 1974, **25**, 546.
2. T. J. Kats and P. J. Gorratt, *J. Am. Chem. Soc.*, 1963, **85**, 2582.
3. A. Tajiri, H. Uchimura and M. Hatano, *Chem. Phys. Lett.*, 1973, **21**, 568.
4. A. Tajiri and M. Hatano, *Bull. Chem. Soc. Jpn*, 1972, **45**, 962.
5. Z. Gasyna, P. N. Schatz, J. P. Hare, T. J. Dennis, H. W. Kroto, R. Taylor and D. R. M. Walton, *Chem. Phys. Lett.*, 1991, **183**, 283.
6. M. Pilch, M. Pawlikowski and O. S. Mortensen, *Chem. Phys.*, 1993, **172**, 277.
7. J. P. Hare, H. W. Kroto and R. Taylor, *Chem. Phys. Lett.*, 1991, **177**, 394.
8. C. H. Henry, S. E. Schatterly and C. P. Slichter, *Phys. Rev. A*, 1965, **137**, 583.
9. (a) J. Mack and M. J. Stillman, in *The Porphyrin Handbook*, Vol. 16, ed. K. Kadish, K. Smith, and R. Guilard, Academic Press, New York, 2003,

p. 43; (b) J. Mack and M. J. Stillman, *Coord. Chem. Rev.*, 2001, **993**, 219–221.

10. V. N. Nemykin, R. G. Hadt, R. V. Belusludov, H. Mizuseki and Y. Kawazoe, *J. Phys. Chem. A*, 2007, **111**, 12901.
11. J. Mack, Y. Asano, N. Kobayashi and M. J. Stillman, *J. Am. Chem. Soc.*, 2005, **127**, 17697.
12. H. Miwa, K. Ishii and N. Kobayashi, *Chem. Eur. J.*, 2004, **10**, 4422.
13. N. Kobayashi and H. Konami, in *Phthalocyanines – Properties and Applications*, Vol. 4, ed. C. C. Leznoff and A. B. P. Lever, VCH, New York, 1996, p. 343.
14. A. Tajiri and Z. Winkler, *Z. Naturforsch. A*, 1983, **38**, 1263.
15. A. Kaito, T. Nozawa, T. Yamamoto and M. Hatano, *Chem. Phys. Lett.*, 1977, **52**, 154.
16. P. C. Minor, M. Gouterman and A. B. P. Lever, *Inorg. Chem.*, 1985, **24**, 1894.
17. (a) J. Mack, S. Kirkby, E. A. Ough and M. J. Stillman, *Inorg. Chem.*, 1992, **31**, 1717; (b) J. Mack and M. J. Stillman, *J. Am. Chem. Soc.*, 1994, **116**, 1292.
18. J. Cheek and J. Dawson, in *The Porphyrin Handbook*, Vol. 7, ed. K. Kadish, K. Smith, and R. Guilard, Academic Press, New York, 2000, p. 339.
19. C. L. Vermeiren, M. Pluym, J. Mack, D. E. Heinrichs and M. J. Stillman, *Biochemistry*, 2006, **45**, 12867.
20. J. C. Grigg, C. L. Vermeiren, D. E. Heinrichs and M. E. P. Murphy, *Mol. Microbiol.*, 2007, **63**, 139.
21. (a) N. Kobayashi, T. Nozawa and M. Hatano, *Bull. Chem. Soc. Jpn*, 1981, **54**, 919; (b) T. Yamamoto, T. Nozawa, N. Kobayashi and M. Hatano, *Bull. Chem. Soc. Jpn*, 1982, **55**, 3059.
22. (a) A. J. Thomson and P. M. A. Gadsby, *Dalton Trans.*, 1990, 1921; (b) P. M. A. Gadsby and A. J. Thomson, *J. Am. Chem. Soc.*, 1990, **112**, 5003.
23. (a) A. J. Thomson, M. K. Johnson, C. Greenwood and P. E. Gooding, *Biochem. J.*, 1981, **193**, 687; (b) M. K. Johnson, A. J. Thomson, A. E. Robinson, K. K. Rao and David O. Hall, *Biochim. Biophys. Acta*, 1981, **667**, 433.
24. P. N. Schatz, R. L. Mowery and E. R. Krausz, *Mol. Phys.*, 1978, **35**, 1537.
25. E. I. Solomon, T. C. Brunold, M. I. Davis, J. N. Kemsley, S.-K. Lee, N. Lehnert, F. Neese, A. J. Skulan, Y.-S. Yang and J. Zhou, *Chem. Rev.*, 2000, **100**, 235.
26. (a) M. Seth, T. Ziegler, A. Banerjee, J. Autschbach, S. J. A. van Gisbergen and E. J. Baerends, *J. Chem. Phys.*, 2004, **120**, 10942; (b) M. Seth, J. Autschbach and T. Ziegler, *J. Chem. Phys.*, 2005, **122**, 094112; (c) M. Seth and T. Ziegler, *J. Chem. Phys.*, 2006, **124**, 144105; (d) M. Krykunov, A. Banerjee, T. Ziegler and J. Autschbach, *J. Chem. Phys.*, 2005, **122**, 074105; (e) M. Seth, J. Autschbach and T. Ziegler, *J. Chem. Theory Comput.*, 2007, **3**, 434; (f) M. Krykunov, M. Seth, T. Ziegler and J. Autschbach, *J. Chem. Phys.*, 2007, **127**, 244102; (g) M. Seth, M.

Krykunov, T. Ziegler, J. Autschbach and A. Banerjee, *J. Chem. Phys.*, 2008, **128**, 144105; (h) M. Seth, M. Krykunov, T. Ziegler and J. Autschbach, *J. Chem. Phys.*, 2008, **128**, 2341025; (i) M. Seth, T. Ziegler and J. Autschbach, *J. Chem. Phys.*, 2008, **129**, 104105.

27. E. Hernandez-Marin, M. Seth and T. Ziegler, *Inorg. Chem.*, 2010, **49**, 6066.
28. G. A. Peralta, M. Seth and T. Ziegler, *Inorg. Chem.*, 2007, **46**, 9111.
29. Z.-L. Cai, M. J. Crossley, J. R. Reimers, R. Kobayashi and R. D. Amos, *J. Phys. Chem. B*, 2006, **110**, 15624.
30. (a) J. Michl, *J. Am. Chem. Soc.*, 1978, **100**, 6801; (b) J. Michl, *Tetrahedron*, 1984, **40**, 3845; (c) J. Michl, *Pure Appl. Chem.*, 1980, **52**, 1549.
31. M. Kirk and K. Peariso, *Polyhedron*, 2004, **23**, 499.

CHAPTER 10

Michl's Perimeter Model in MCD Spectroscopy

One of the factors that can make magnetic circular dichroism (MCD) spectroscopy somewhat inaccessible to many organic and inorganic chemists is that ligand field theory and/or molecular orbital theory is usually required to interpret the spectral data. Although valence bond theory is still widely used as the conceptual framework for visualising bonding and reactivity, a molecular orbital (MO) approach is required to account for the aromatic properties of benzene and other aromatic and heteroaromatic π-systems. This chapter will focus on the use of molecular orbital theory to analyse the MCD spectra of aromatic and antiaromatic cyclic polyenes. Since the electron density for each molecular orbital is spread across the entire complex, the nodes and antinodes associated with the orbital angular motion around a π-system can be observed on the cyclic perimeters of aromatic molecules, Figure 10.1. In the context of benzene, linear combinations of the six carbon $2p_z$ orbitals, which combine within the molecular orbital approach, result in three bonding and three antibonding MOs with an $M_L = 0, \pm 1, \pm 2, 3$ sequence in ascending energy terms. The resonance energy of benzene, which accounts for the lack of reactivity, is based on the fact that all of the stabilised bonding MOs are fully occupied. This forms the basis of Hückel's $4N + 2$ rule for aromaticity.

In the late 1940s, Platt[1] proposed a free electron model for describing the energies of the π-system perimeter MOs in wavenumbers (cm^{-1}), based on an orbital angular momentum quantum number (essentially the same as M_L) for the cyclic perimeter, which was referred to as q:

$$E = \frac{q^2 \mathrm{h}^2}{2\mathrm{m_e}\mathrm{l}^2} \qquad (10.1)$$

where $\mathrm{m_e}$ is the mass of the electron, h is Planck's constant and l is the length of the cyclic perimeter (Å). Since the HOMO and LUMO of benzene have

Circular Dichroism and Magnetic Circular Dichroism Spectroscopy for Organic Chemists
By Nagao Kobayashi, Atsuya Muranaka and John Mack

Published by the Royal Society of Chemistry, www.rsc.org

$M_L = \pm 1$ and ± 2 nodal properties, respectively, Figure 10.1, the $1e_{1g}$ and $1e_{2u}$* MOs can be viewed as being linked by an allowed $\Delta M_L = \pm 1$ and a forbidden $\Delta M_L = \pm 3$ transition, Figure 10.2. Platt[1] referred to these as the B and L bands, respectively. The forbidden $\Delta M_L = \pm 3$ transitions lie at lower energy, on the basis of Hund's second rule that for a given multiplicity the lower energy

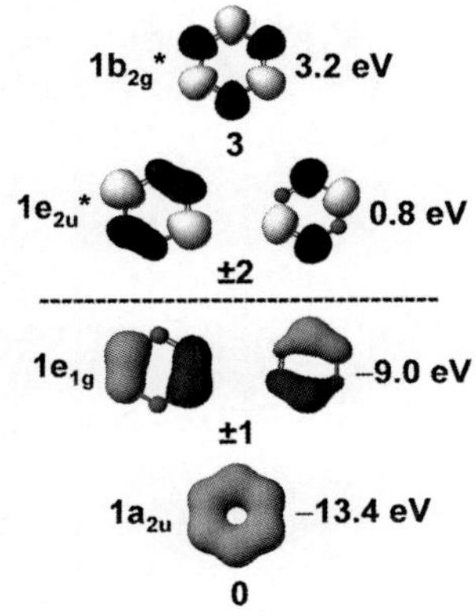

Figure 10.1 The real π-MOs of benzene at an isosurface of 0.05 a.u. based on a density functional theory (DFT) geometry optimisation. Nodal patterns are observed, due to the $M_L = 0$, ± 1, ± 2 and 3 properties associated with the corresponding complex wavefunctions.

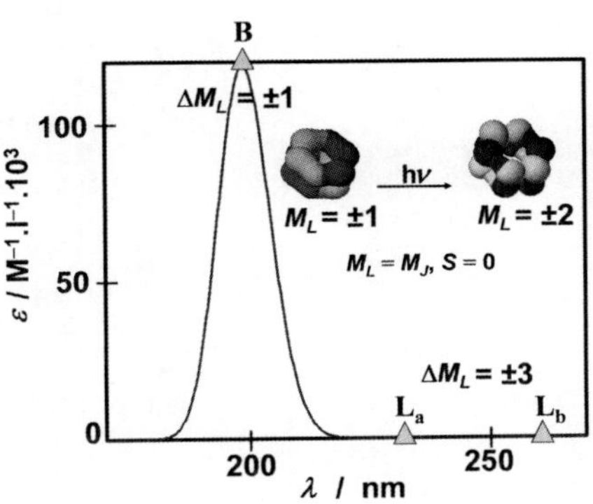

Figure 10.2 Calculated UV-visible spectrum of benzene based on a DFT geometry optimisation and INDO/s calculation. Triangles denote band centres associated with spin allowed electronic transitions. The B excited state is orbitally degenerate, $^1E_{1u}$. In contrast, the L_a and L_b excited states arising from the $\Delta M_L = \pm 3$ transitions are nondegenerate with $^1B_{1u}$ and $^1B_{2u}$ symmetry in the case of benzene due to the D_{6h} symmetry. The L_a state is associated with nodes through the bonds, while the L_b state is associated with nodes through the atoms, Figure 10.1.

state has the greater angular momentum, Figure 10.2. The B band is observed at 180 nm and is strongly allowed ($\varepsilon = 65\,000$), while the forbidden L bands lie at 200 nm ($\varepsilon = 8000$) and at 254 nm ($\varepsilon = 254$) gaining intensity by vibrational borrowing.

Moffitt[2] subsequently developed a more complex linear combination of atomic orbitals (LCAO) based model of the cyclic perimeter and noted the fact that, in the case of cyclic heteroaromatic polyenes, the nodal patterns for structural homologues derived from the same basic ideal charged or uncharged parent C_nH_n perimeter are consistently retained and that the molecular orbital method used to predict the energies and relative intensities of the major spectral bands can be simplified on this basis. A 16 atom 18 π-electron $C_{16}H_{16}{}^{2-}$ π-system, consistent with Hückel's $4N + 2$ rule for aromaticity, can be viewed as being the parent cyclic perimeter of metal porphyrinoids based on a consideration of the inner ligand perimeter. An $M_L = 0, \pm 1, \pm 2, \pm 3, \pm 4, \pm 5, \pm 6, \pm 7$ and 8 sequence is, therefore, observed in the molecular orbitals predicted for both $C_{16}H_{16}{}^{2-}$ and zinc porphyrin on the basis of DFT optimisations. In the case of metal porphyrin complexes, the symmetry of the ligand is lowered from D_{16h} to D_{4h} owing to the introduction of four pyrrole rings, but the same basic nodal pattern sequence can still be observed clearly in the four frontier π-MOs, Figure 10.3, resulting in a similar calculated UV-visible spectrum, Figure 10.4, to that obtained for benzene. The basis of the perimeter model approach to the electronic structure of porphyrinoids is that, although the original degenerate HOMO and LUMO orbitals, $1e_{4u}$ and $1e_{5g}{}^*$, may and

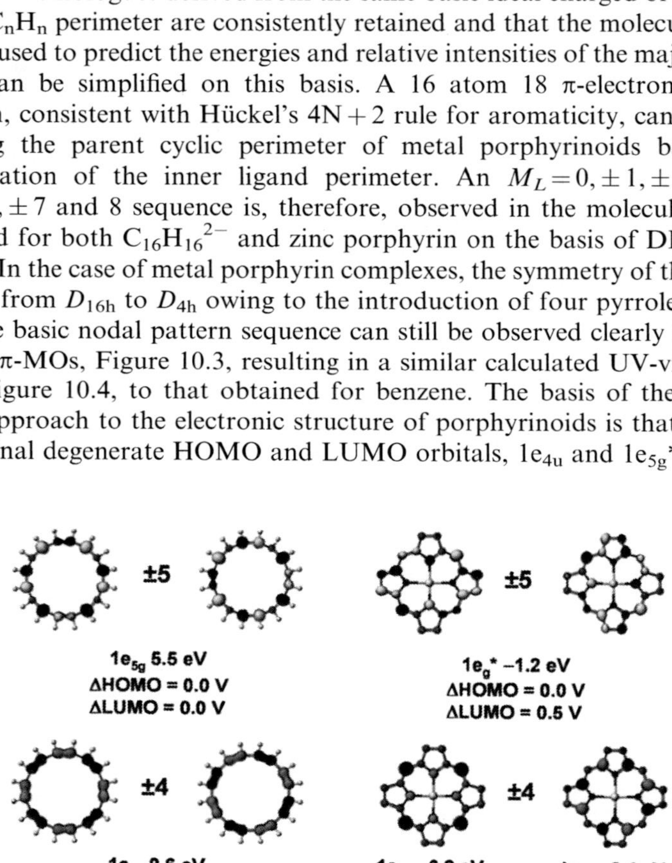

Figure 10.3 The real π-MOs of the D_{16h} symmetry $C_{16}H_{16}{}^{2-}$ species (left) and the D_{4h} symmetry zinc porphyrin complex (right) at an isosurface of 0.05 a.u., based on a DFT geometry optimisation. Nodal patterns can be observed on the inner cyclic perimeter based on the magnetic quantum number associated with the corresponding complex wavefunctions with $M_L = 0, \pm 1, \pm 2, \pm 3, \pm 4, \pm 5, \pm 6, \pm 7, 8$ properties. Only the four frontier orbitals that form the basis of Michl's perimeter model are shown. The ΔHOMO and ΔLUMO terminology used by Michl[3] relates to the perturbation of the frontier π-MOs relative to the $1e_{4u}$ and $1e_{5g}$ MOs of the $C_{16}H_{16}{}^{2-}$ parent hydrocarbon cyclic perimeter, respectively.

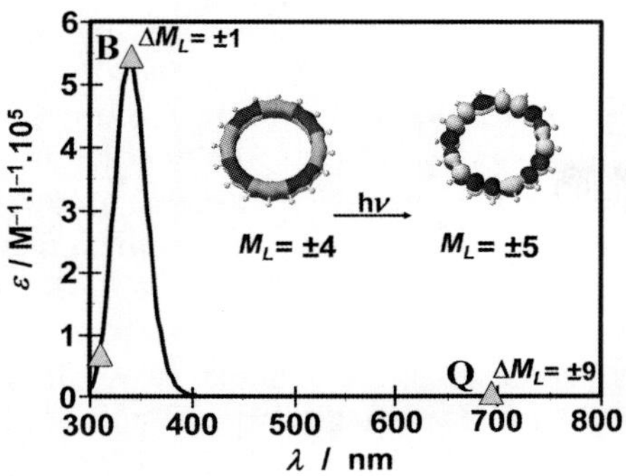

Figure 10.4 Calculated UV-visible spectrum of a D_{16h} symmetry $C_{16}H_{16}^{2-}$ perimeter based on a DFT geometry optimisation and an INDO/s calculation. Triangles denote band centres associated with spin allowed transitions. In the case of the $C_{16}H_{16}^{2-}$ perimeter the allowed and forbidden states are both orbitally degenerate with $^1E_{1u}$ and $^1E_{7u}$ symmetry, respectively. Michl refers to these bands as the B and L bands.[3]

do split in energy terms when the symmetry is lowered from the D_{16h} of $C_{16}H_{16}^{2-}$, the four frontier π-MOs still retain the same set of nodal properties. The electronic structures of low symmetry cyclic polyenes can, therefore, be described in terms of perturbations to the structure of the parent hydrocarbon, which modify the magnetic properties associated with the π-system in a manner that can be readily probed by MCD spectroscopy. In the 1970s, Michl[3] developed a general theory on this basis to account for the MCD spectroscopy of aromatic and heteroaromatic cyclic π-electron systems, which can be used to predict the sign sequences of $\mathcal{A}_1$ (or of coupled oppositely signed $\mathcal{B}_0$) terms on the basis of a consideration of the excited state magnetic moments induced by Platt's forbidden and allowed L and B transitions[1] and of how various structural perturbations modify the relative energies of the four frontier π-MOs, which are derived from the HOMO and LUMO of the parent perimeter, Figure 10.5. It was hoped at the time that this would result in the MCD technique becoming a more widely used spectroscopic tool within organic chemistry but this largely failed to happen. Although it has a much wider potential applicability, Michl's perimeter model has tended subsequently to be used primarily to provide a conceptual framework for rationalising the effects of different symmetry lowering perturbations on the electronic structures and optical properties of porphyrinoids.[4]

10.1 Michl's 4N + 2-Perimeter Model for Aromatic π-Systems

Since the π-system forms a cyclic perimeter comprised of $2p_z$ atomic orbitals of carbon and nitrogen atoms, complex wavefunctions permit a description of the net electron circulation upon electronic excitation based on the component atoms, which is not possible on the basis of the more commonly used real wavefunctions. A stack of MOs can be constructed for the $C_{16}H_{16}^{2-}$ parent perimeter of conventional radially symmetric tetrapyrrole porphyrinoids,[3] Figure 10.5. For each set of degenerate complex wavefunctions Ψ_N and Ψ_{-N}, the Ψ_{-N} MO involves electron circulation (based on a redistribution of electronic charge from one $2p_z$ atomic orbital to another) in a clockwise direction from the lowest energy $M_L = 0$ MO, while the Ψ_N MO involves electron circulation in an anticlockwise direction. When a magnetic field is applied parallel to the axis of light propagation, absorption of a photon of left circularly polarised (lcp) and right circularly polarised (rcp) results in either a $\Delta M_L = +1$ or a $\Delta M_L = -1$ transition, since the electric vector of the incident electromagnetic wave rotates clockwise or anticlockwise once per wavelength and the total angular momentum within the system must be conserved. There is either clockwise or anticlockwise charge circulation on the perimeter of the $\pi\pi^*$ excited state, based on the direction of rotation of the electric vector of the incident photon, Figure 10.5.

In high symmetry cyclic polyenes, such as benzene, in which the π-MOs associated with the perimeter retain orbital degeneracy, the orbital angular momentum (OAM) is no longer fully compensated by the motion of a second electron in the MO of the opposing handedness when an electron is promoted to the LUMO in an electric dipole allowed manner ($\Psi_N \rightarrow \Psi_{N+1}$ in the case of lcp and $\Psi_{-N} \rightarrow \Psi_{-N-1}$ in the case of rcp). When OAM is greater in the LUMO than in the HOMO, a negatively charged particle is effectively formed on the perimeter. The alignment of the magnetic dipole induced within the excited state can be visualised on the basis of Ampere's rule for current flow in a

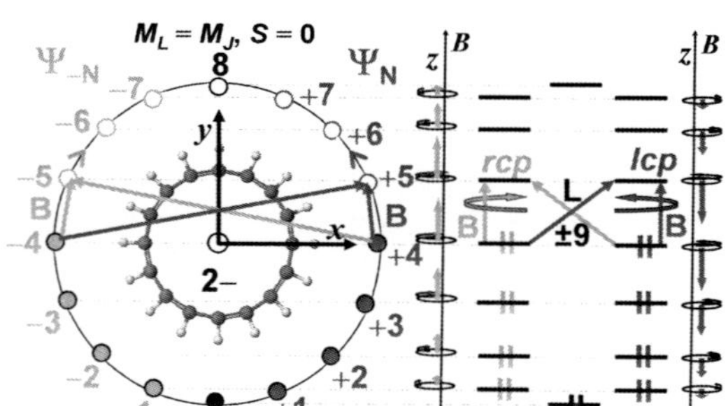

Figure 10.5 Michl's perimeter model as applied to $C_{16}H_{16}^{2-}$ (left).[4] The circle is a diagrammatic representation of the clockwise (light grey) and anticlockwise (dark grey) motion of π-system electrons on the $C_{16}H_{16}^{2-}$ parent perimeter of porphyrinoid compounds based on the sixteen $2p_z$ atomic orbitals of the inner ligand perimeter, generating the $M_L = 0, \pm 1, \pm 2, \pm 3, \pm 4, \pm 5, \pm 6, \pm 7, 8$ value for each complex wavefunction MO. The perspective is towards the light source. The magnitudes of the magnetic moments induced in the applied field within each MO (right) are plotted diagrammatically, based on the results of a set of LCAO calculations carried out by Michl.[3c]

solenoid. When the fingers of the left hand are curled around the loop (*i.e.* the convention would normally be to use the right hand for conventional current flow with a positive charge carrier), the thumb points in the direction of the induced dipole moment. A positive magnetic dipole moment, which points along the *z*-axis of the applied field, is the energetically favoured alignment within the Zeeman splitting of states. Since the perspective is towards the detector rather than the source, this corresponds to circulation induced by the absorption of a photon of incident rcp. This results in a derivative-shaped Faraday $\mathcal{A}_1$ term signal with a −, + sign sequence. This is referred to as a positive $\mathcal{A}_1$ term because the higher energy component has a positive ΔA_{l-r} signal. Positive Faraday $\mathcal{A}_1$ terms, Figure 10.6, are usually the dominant spectral feature of high symmetry cyclic polyenes, such as radially symmetric porphyrinoids, bcause the OAM associated with circulation of charge in the LUMO level is usually greater than that associated with circulation in the HOMO level.

The induced ππ* magnetic moments for Platt's electronically allowed B and forbidden L transitions for benzene (which would have $\Delta M_L = \pm 1$ and $\Delta M_L = \pm 3$ properties in a free electron model approach) were referred to by Michl as the μ^- and μ^+ moments, respectively. When the symmetry is lowered from that of the parent perimeter to form heteroaromatic compounds, the OAM properties of the π-system are modified. For example when the symmetry of a D_{16h} symmetry $C_{16}H_{16}^{2-}$ species is modified by incorporating pyrrole moieties to form a D_{4h} symmetry porphyrinoid, there is a lifting of the degeneracy of the HOMO of the parent perimeter and this results in a mixing of the allowed and forbidden properties of Q and B bands. It should be noted that the L band of Michl's 4N + 2 perimeter model is usually referred to as the Q band in the context of porphyrinoids. Michl demonstrated that the degree of mixing, α, and the magnitude of the induced magnetic dipole moments of the Q and B excited states ($\mu_{(Q)}$, $\mu_{(B)}$) can be estimated using equations (10.2–10.4), which are based on the observed dipole strengths, $\mathcal{D}_0$, for the Q and B bands, which can either be derived empirically by spectral band deconvolution, or from TD-DFT calculations by converting the predicted oscillator strength values.

$$\mathcal{D}_0(Q)/\mathcal{D}_0(B) = \tan^2\alpha \tag{10.2}$$

$$\mu_{(B)} = \pm(\cos^2\alpha)\mu^- \pm (\sin^2\alpha)\mu^+ \tag{10.3}$$

$$\mu_{(Q)} = \pm(\sin^2\alpha)\mu^- \pm (\cos^2\alpha)\mu^+ \tag{10.4}$$

When the symmetry of the parent perimeter is perturbed still further so that the four-fold axis of symmetry of radially symmetrical porphyrinoids is removed, the MCD spectrum is completely dominated by Gaussian-shaped $\mathcal{B}_0$ terms. Michl demonstrated successfully that the alignment of the magnetic moments, and hence the sign sequence in the MCD spectrum, is determined by

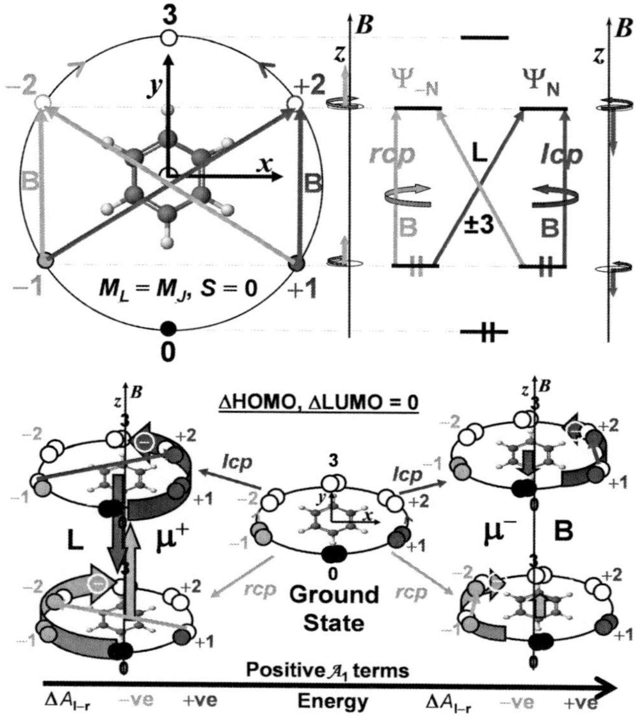

Figure 10.6 The sign orderings in ascending energy terms of the positive $\mathcal{A}_1$ terms associated with Platt's B and L bands[1]) (LEFT) is related to Ampere's rule (right-hand rule) for conventional current flow within a solenoid (RIGHT) in which the thumb is aligned with the magnetic dipole induced in the core of the current loop, when the fingers of the right hand are curled around the loop in the direction of conventional current flow. The left hand is used in this instance, since the charge carrier is negatively charged. The occupancy of the six complex MOs within Michl's perimeter model,[3]) Figure 10.5, and the alignment of the μ^+ and μ^- magnetic dipoles, associated with the forbidden L and allowed B transitions, induced through the absorption of a photon of either lcp or rcp are indicated with shaded circles and arrows, respectively, using the same scheme adopted in Figure 10.5 for the Q and B bands in Gouterman's 4-orbital model. The resulting MCD sign sequence in ascending energy terms is shown based on the Zeeman splitting of states, which arises from the fact that a moment aligned with the applied field is energetically favoured.

whether the OAM of the incident lcp or rcp photon is conserved by the circulation of the positively charged hole left in the HOMO level or the negatively charged particle circulating in the LUMO level, based on the larger μ^+ moment associated with the forbidden L transition. In the context of porphyrinoids this is determined by the relative energy separations of the frontier π-MOs derived from the $1e_{4u}$ HOMO and $1e_{5g}$ LUMO of the $C_{16}H_{16}^{2-}$ parent hydrocarbon perimeter (Michl's terminology for this is the ΔHOMO, ΔLUMO value).

When ΔHOMO >> ΔLUMO, the OAM associated with the excited electron predominates and there is a −, +, −, + ΔA_{l-r} intensity pattern with ascending energy for the Faraday $\mathcal{B}_0$ terms associated with the Q and B bands. When ΔLUMO >> ΔHOMO circulation of the positive hole left within the HOMO becomes the predominant factor. The magnetic moments induced by the absorption of a photon of lcp and rcp are reversed and the ΔA_{l-r} sign sequence for the Faraday $\mathcal{B}_0$ terms becomes +, −, +, −. In the context of low symmetry compounds, the key to applying Michl's perimeter model is to analyse the effect of different structural perturbations such as the introduction of heteroatoms, fused rings and peripheral substituents on the relative energies of the four frontier π-MOs and how the ΔHOMO and ΔLUMO values and the magnitude of the HOMO–LUMO band gap are based on a consideration of the alignment of the nodal planes. Changes observed in the signs of the Faraday $\mathcal{B}_0$ terms associated with the L and B transitions due to the relative magnitudes of the ΔHOMO and ΔLUMO values can be used to validate semi-empirical and DFT descriptions of the electronic structure.[4]

10.2 Michl's 4N-Perimeter Model for Antiaromatic π-Systems

In the late 1990s, Michl and coworkers[5] developed a 4N-perimeter model to analyse the electronic spectra of antiaromatic species, and they reported subsequently the analyses of several small cyclic π-electron systems on this basis.[6–10] Six frontier π-MOs derived from the HOMO, SOMO and LUMO of the parent hydrocarbon perimeter (referred to by Michl as h−, h+, s−, s+, l− and l+) play the key role in the low-energy singlet electronic states, Figure 10.7. This is in marked contrast with the (4N+2)-electron perimeter model where the L and B bands arise from transitions between the four frontier π-MOs derived from the HOMO and LUMO of the parent hydrocarbon perimeter.[3]

Michl *et al.* used new terminology to describe even-electron cyclic 4N-π-systems.[5] The term "antiaromatic" is reserved for π-systems that have very strong biradical-like character, while 4N-electron perimeters in which the structural perturbations are strong enough to remove the biradical-like character are described as "unaromatic". The lowest-energy band is associated with the s− → s+ transition, Figure 10.7, and is referred to as the S band. Since the transition is of an intrashell nature, it is electric dipole forbidden. Although the S band is magnetic dipole allowed, the absorption and hence also the MCD intensity is predicted to be zero.[7] The other four electronic transitions are intershell, being derived from HOMO → SOMO and SOMO → LUMO one-electron transitions, Figure 10.7. Two weakly (N_1, N_2) and two strongly (P_1, P_2) electric-dipole allowed transitions are predicted. The N and P transitions are somewhat analogous to the L and B transitions of aromatic systems. Configuration interaction between states arising from the h+ → s+ and s− → l− transitions causes cancellation and intensification of the transition dipole

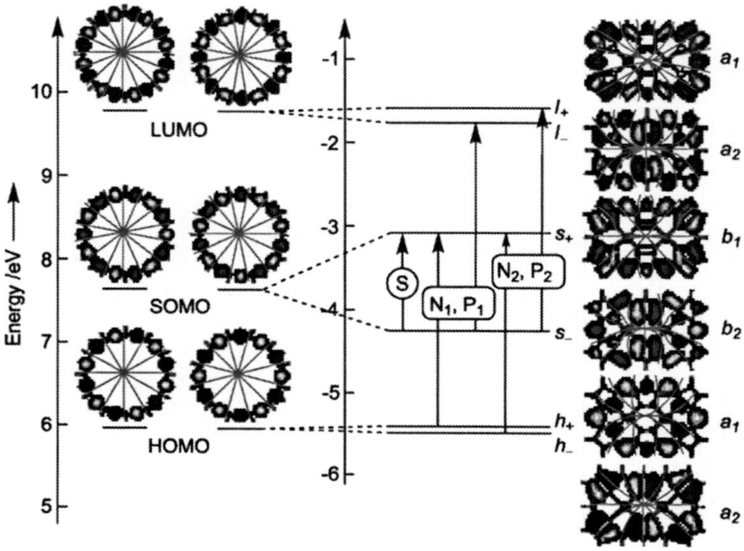

Figure 10.7 Isosurface plots of the frontier π-MOs, energy levels and perimeter labels of an antiaromatic $C_{24}H_{24}^{4-}$ cyclic perimeter and an "unaromatic" Au^{III} hexaphyrin species.[16] Arbitrary nodal lines are drawn on the isosurface plots. HOMO = s−, LUMO = s+.

moments, resulting in weak and intense N^1 and P^1 bands, respectively. Weak N_2 and intense P_2 bands arise from the h− → s+ and s− → l+ transitions in a similar manner, Figure 10.7.

10.3 Applications of Michl's Perimeter Model

10.3.1 Negative Faraday $\mathcal{A}_1$ MCD Term in a Cyclobutadiene Dianion[11]

Kobayashi and coworkers[11] analysed the MCD spectroscopy of a 4 atom 6 π-electron cyclobutadiene dianion (CBD^{2-}) to test the validity of Michl's perimeter model.[3] CBD^{2-} has attracted attention from both synthetic and theoretical viewpoints. Based on Hückel's rule, the presence of six π-electrons would be expected to result in an aromatic molecule with a planar structure. It was pointed out, however, that CBD^{2-} derivatives could adopt distorted structures, owing to electronic repulsion associated with the large negative charge density on each atom. When the dilithium salt of the *tetrakis*(trimethylsilyl)cyclobutadiene dianion, Figure 10.8, inset, was prepared, its X-ray crystallographic analysis demonstrated that the four-membered ring is planar with approximate D_{4h} symmetry, and the optical properties of CBD^{2-} were examined by MCD spectroscopy to test the validity of Michl's perimeter

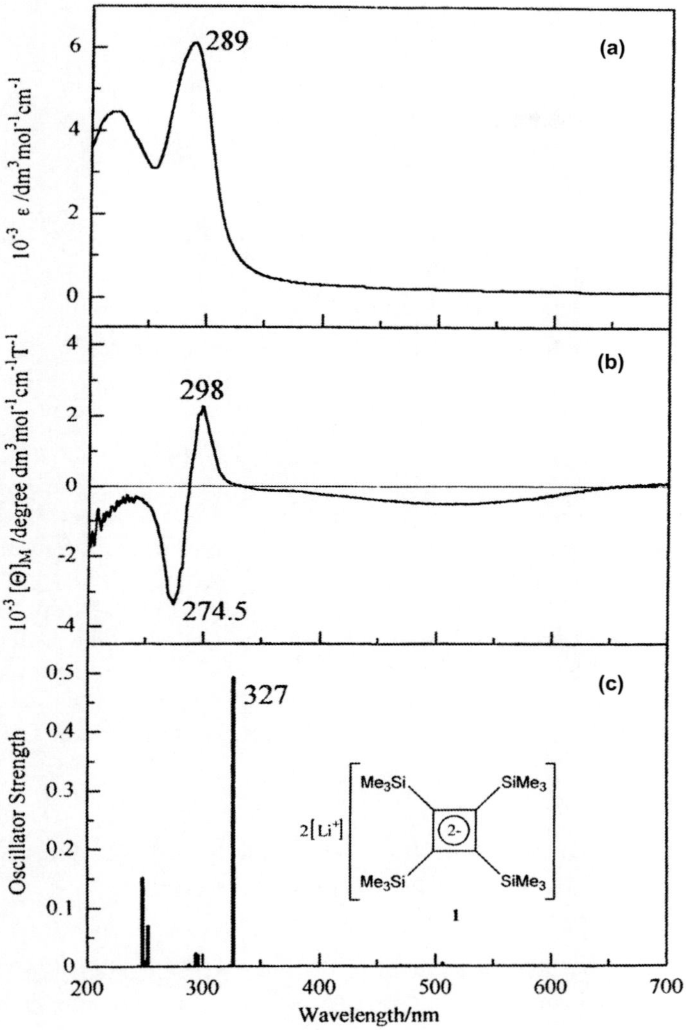

Figure 10.8 Electronic absorption (a) and MCD (b) spectra of the dilithium salt of *tetrakis*(trimethylsilyl)cyclobutadiene dianion with the calculated (ZINDO/S) spectrum (c).[11]

model in this context, Figure 10.8. An MO calculation predicts that the aromatic CBD^{2-} has four π-MOs (a_{2u}, e_g and b_{2u}), as shown in Figure 10.9, resulting in an orbitally degenerate HOMO and a nondegenerate LUMO. In a four-membered ring with D_{4h} symmetry, the groundstate is $^1A_{1g}$ and an $e_g \rightarrow b_{2u}$ excitation leads to a ^1E_u excited state. Since $^1A_{1g} \rightarrow {}^1E_u$ transitions are electric dipole allowed with x/y-polarisation, only one intense derivative-shaped Faraday $\mathcal{A}_1$ term is expected in the MCD spectrum. A negative Faraday

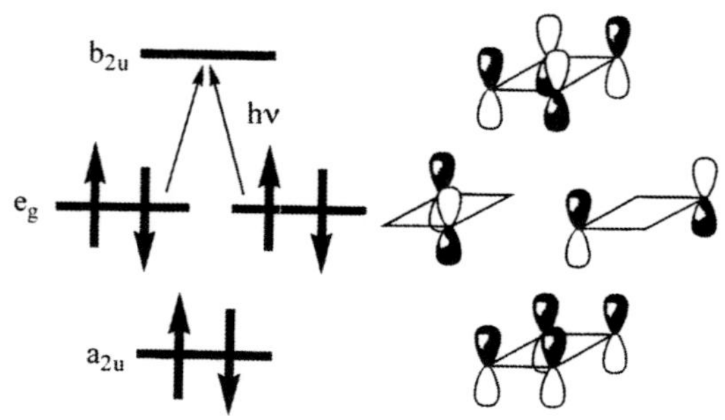

Figure 10.9 The π-MOs of the aromatic cyclobutadiene dianion.[11]

$\mathcal{A}_1$ term is observed at 289 nm for the $\Psi_{\pm 1} \rightarrow \Psi_2$ transition as would be anticipated on the basis of Michl's perimeter model, Figure 10.10. Since there is no orbital angular momentum associated with the LUMO, the positive hole within the π-system HOMO is the dominant factor in conserving the orbital angular momentum of the incident photon in charge circulation terms, and the alignment of the magnetic dipole moments is reversed from that normally anticipated. Given that the microstate associated with the absorption of lcp is stabilised by the Zeeman splitting of states, the negative lobe of the $\mathcal{A}_1$ term intensity lies at higher energy rather than the −, + sign sequence pattern that is normally observed in ascending energy terms when the orbital angular momentum is conserved by the electron circulating in the LUMO level, Figure 10.10.

10.3.2 The Effect of Ligand Non-planarity on the Alignment of the Excited State Magnetic Moments of Zinc Tetraphenyltetraacenaphthoporphyrin

When bulky substituents are added to the porphyrin ligand at both the *meso*-carbons and the β-carbons of the pyrrole moieties, the ligand tends to fold as a result of the steric crowding.[12] In most cases the main C_4 axis of symmetry is replaced by an S_4 symmetry axis due to saddling of the π-system. Since the degeneracy of the LUMO is retained on moving from D_{4h} to D_{2d} symmetry, $\mathcal{A}_1$ terms are still anticipated for the Q and B bands in the MCD spectra. The MCD spectra of M^{II}TPTANP complexes of main group and closed shell d^{10} metals were found to exhibit anomalous negative Faraday $\mathcal{A}_1$ terms (*i.e.* a +, − sign sequence in ascending energy terms), while the normal −, + sign sequence was observed for free base compounds and dication species.[13] Michl referred to cyclic polyenes such as TPTANPs in which ΔHOMO, ΔLUMO ≈ 0 as soft MCD chromophores,[3] since small structural modifications can potentially reverse the alignment of the induced magnetic moments of the ππ* excited states and hence the sign sequence observed for the Faraday $\mathcal{A}_1$ terms or coupled oppositely signed Faraday $\mathcal{B}_0$ terms associated with the main Q(0,0)

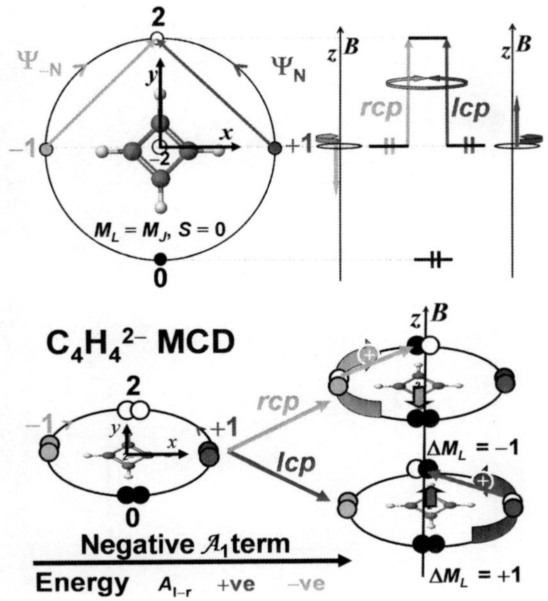

Figure 10.10 The origin of negative Faraday $\mathcal{A}_1$ terms is based on a reversal in the alignment of the excited state magnetic dipoles induced by the Zeeman splittings of states. In an analogous manner to the $M_L = 0$ MO, the orbitally nondegenerate $M_L = 2$ MO has no net orbital angular momentum; the lcp and rcp transitions, therefore, both obey the $\Delta M_L = \pm 1$ selection rule. The occupancy of the four complex MOs within Michl's perimeter model,[3] Figure 10.1, and the alignment of the magnetic dipoles induced through the absorption of a photon of either lcp or rcp are indicated with dark and light grey circles and arrows, respectively (bottom). The resulting MCD sign sequence in ascending energy terms is shown, based on the fact that the moment aligned with the applied field is energetically favoured.

and B(0,0) bands, see Figure 10.4. The saddling of the ligand is believed to quench the orbital angular momentum properties of the LUMO level to a greater extent than those of the HOMO level owing to there being larger MO coefficients on the β-carbons of the pyrrole moieties and on the fused acenaphthalene rings. The orbital angular momentum of the incident photon is, therefore, conserved by the circulation of the positively charged hole left in the HOMO, which reverses the alignment of the excited state magnetic moments and results in negative Faraday $\mathcal{A}_1$ terms.

10.3.3 The Application of the 4N and 4N+2 Perimeter Model to the Spectra of Hexaphyrin(1.1.1.1.1.1)s

The synthesis and characterisation of *meso-hexakis*(pentafluorophenyl)-substituted hexaphyrin(1.1.1.1.1.1)s with multiple gold-carbon bonds has been

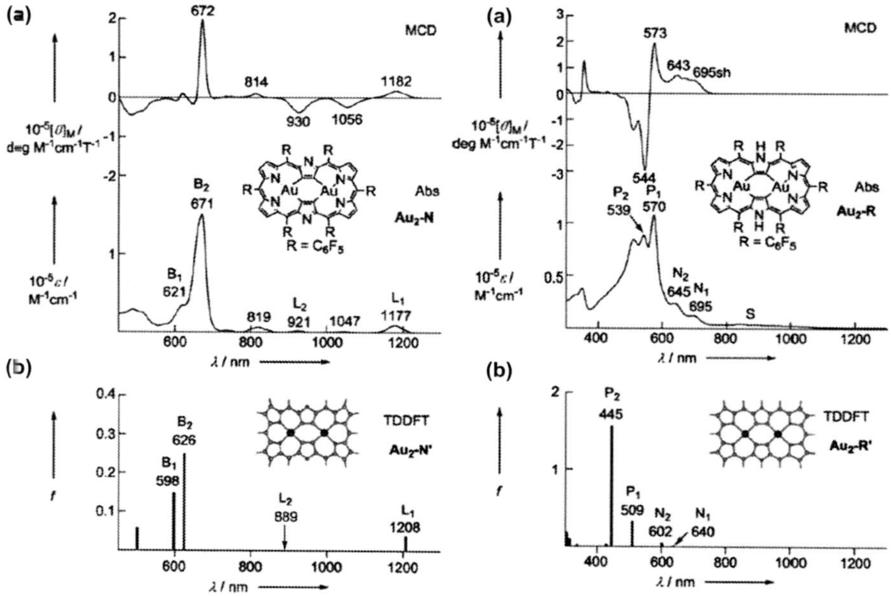

Figure 10.11 Electronic absorption and MCD spectra of aromatic (left) and "unaromatic" (right) Au^{III} hexaphyrin species (Au_2-N and Au_2-R).[16] The main electronic bands are assigned on the basis of the sign sequences in the MCD spectra, TD-DFT calculations of model complexes and on Michl's $(4N+2)$[3] and 4N[5,6] perimeter models.

reported recently,[14] Figure 10.11. Two-electron reduction of the aromatic Au^{III} hexaphyrins (Au_2-N) leads to reduced species with remarkable paratropic ring current properties (Au_2-R). The neutral and reduced forms can therefore be regarded as 26- and 28-π-electron systems. Upon reduction, there are marked changes in the electronic absorption spectrum of the hexaphyrin complex. Several NIR absorption bands observed within the Au_2-N spectrum are not present in the spectrum of Au_2-R, Figure 10.11. Similar spectra were also observed when various peripheral substituents were added at the pyrrole β-positions.[15] An analysis[16] of the MCD spectra was used to validate the results of TD-DFT calculations. The near IR region of the MCD spectrum, Figure 10.11, of Au_2-N contains two overlapping pairs of coupled and oppositely signed Faraday $\mathcal{B}_0$ terms at 1182 and 930 nm and at 1058 and 814 nm, which can be assigned as L(0,0) and L(0,1) bands, while bands at 672 and 621 nm can be assigned to the B transition on the basis of the $4N+2$ model,[3] see Figure 10.5. In contrast, almost no MCD intensity is observed for the low energy tail of absorbance between 800 and 1100 nm in the spectrum of Au_2-R, which is the pattern that would be anticipated for a band associated with the forbidden intrashell S transition in the context of the 4N – model,[5,6] see Figure 10.7. The Faraday $\mathcal{B}_0$ terms at 544, 573, 645 and 695 nm can be readily assigned as P_2, P_1, N_2 and N_1 bands, respectively. In this context, Michl's perimeter model provides a readily accessible conceptual framework for rationalising the key trends in what would otherwise have appeared to be highly complex spectra.

10.3.4 Using Michl's Perimeter Model to Conceptualise the Optical Properties of Low Symmetry Aromatic π-Systems

While it is now relatively straightforward for researchers to carry out TD-DFT calculations routinely for aromatic and heteroaromatic π-systems, it is important to bear in mind that much of organic and inorganic chemistry revolves around an ability to predict properties on the basis of an understanding of how aspects of structure and bonding modify molecular properties. It is, therefore, still useful to be able to deduce the relative size of the ΔHOMO and ΔLUMO values from Michl's perimeter model using a qualitative back of an envelope approach, since this makes it possible to conceptualise key aspects of the electronic structure of aromatic compounds and to predict trends in their optical and redox properties. Michl[3b] referred to the two frontier MOs derived from the HOMO and LUMO of the parent perimeter in which nodal planes lie on the *y*-axis as **a** and **–a**, respectively, Figure 10.12, while the corresponding MOs that lie on antinodes are referred to as **s** and **–s**. This nomenclature simplifies the comparison of structurally related molecules of different symmetry considerably. Once the alignment of the nodes and antinodes of the π-system is firmly established, the effect of structural perturbations can be predicted readily on the basis of factors such as whether the perturbation is introduced at a node or an antinode of the parent perimeter.

Figure 10.12 contains the nodal properties of the frontier MOs of the $C_{16}H_{16}^{2-}$ anion and illustrates their interactions with what can be viewed as the MOs of four C_2H_2 bridging units, which when combined with the incorporation of the four pyrrole nitrogens results in the formation of a porphyrin dianion.[17] The D_{4h} symmetry and $M_L = \pm 5$ nodal patterns dictate that the **–a** and **–s** MOs are degenerate with nodal patterns that differ along the *x*- and *y*-axes and are rotated by 90° from each other. The **a** and **s** MOs of the porphyrin dianion are accidentally near degenerate despite the fact that the four-fold radial symmetry of the nodal planes would normally be expected to result in a significant energy splitting in the context of D_{4h} symmetry, since there are nodes and antinodes on alternating sets of atoms. The energy of the **s** MO is lowered by the introduction of the pyrrole nitrogen heteroatoms, while the C_2H_2 bridges are attached at points on the 16-atom inner ligand perimeter where the **s** MO has nodal planes and, therefore, do not modify its energy significantly. In the **a** MO, the energy is lowered as a result of a strong bonding interaction with what can be viewed as four peripheral C_2H_2 antibonding MOs, while in contrast the pyrrole nitrogens lie on nodal planes and, therefore, have no effect. The effects of the structural perturbations that are introduced, therefore, effectively balance each other.

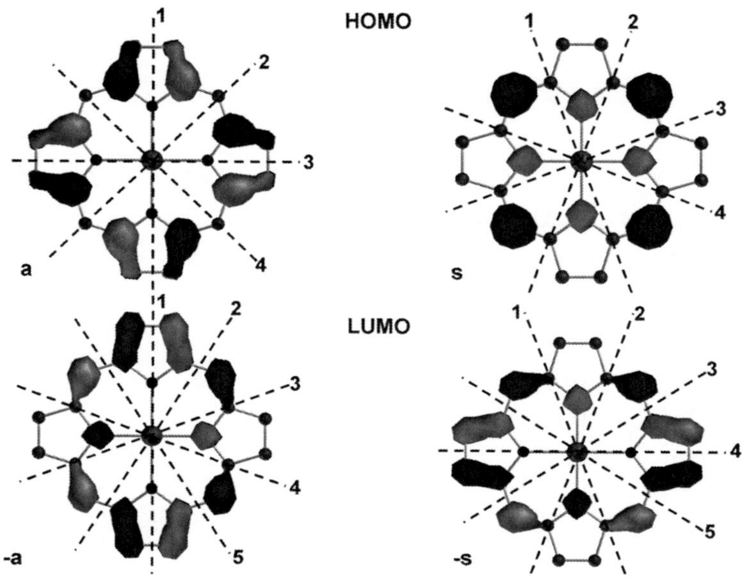

Figure 10.12 Michl[3] introduced **a**, **s**, –**a** and –**s** nomenclature to describe the four frontier π-MOs of aromatic and heteroaromatic π-systems, which are derived from the doubly degenerate HOMO and LUMO of a high symmetry parent hydrocarbon perimeter, based on whether there is a nodal plane (**a** and –**a**) or an antinode (**s** and –**s**) at the position which corresponds to the $M_L = 0$ in Figure 10.5. In this instance, the four frontier π-MOs of zinc porphyrin have an isosurface value of 0.04 a.u.; that is, one of the pyrrole nitrogens on the *y*-axis. Once the alignment of the nodal planes has been defined clearly on the basis of a consideration of molecular symmetry, the effect of different structural perturbations such as heteroatom substitution, bridging and crosslinks can be readily conceptualised on a qualitative basis through a consideration of the relative size of the MO coefficients on each atom on the perimeter.

When hydrogenated chlorin derivatives are formed as a result of the reduction of the peripheral bonds of the pyrrole moieties, the degeneracy of the HOMO and/or LUMO levels is broken significantly. For instance, in bacteriochlorin, **2**, the ΔLUMO value is greatly increased because the interaction between the inner perimeter and the peripheral bonds differs along the *x*- and *y*-axes, see Figure 10.13. At the same time, the increase in the ΔHOMO value is less significant, because no change occurs in the energy of the **s** MO owing to the presence of nodal planes on the α-carbons of the pyrrole moieties, which eliminate any significant interaction with the peripheral β-carbons. In contrast, in the context of isobacteriochlorin, **3**, the ΔHOMO value is the same as that of the bacteriochlorin because the nodal patterns of **a** and **s** MOs are radially symmetrical, but ΔLUMO ≈ 0, since the perturbation to the structure of the parent perimeter is identical along the *x*- and *y*-axes, see Figure 10.13. Djerassi and coworkers[18] subsequently used a combination of MCD

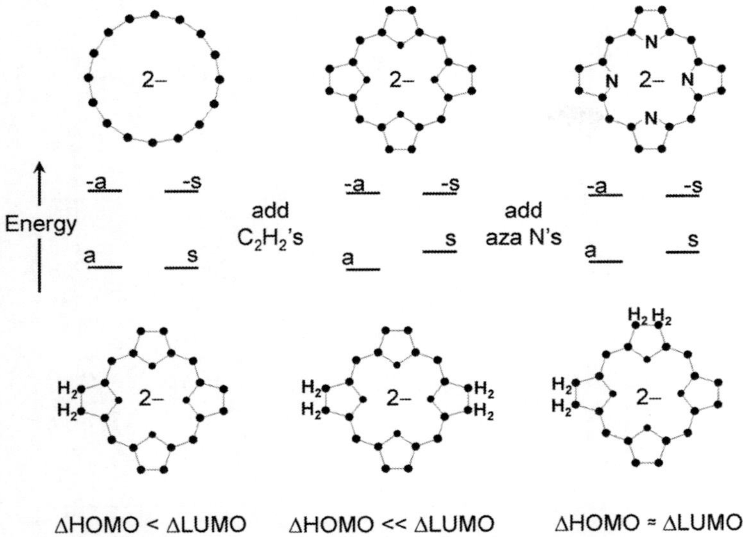

Figure 10.13 The effect on the relative energies of the **a**, **s**, **–a** and **–s** MOs of introducing C_2H_2 bridges and nitrogen heteroatoms to form the porphyrin dianion from a D_{16h} symmetry $C_{16}H_{16}^{2-}$ parent perimeter.[17] The relative size of the ΔHOMO and ΔLUMO values of chlorin compounds can be readily predicted even on a qualitative basis on the basis of the alignment of the nodal planes in Figure 10.12.

spectroscopy and INDO/s calculations to study the spectroscopy of structurally related porphyrins, chlorins, bacteriochlorins and isobacteriochlorins. This research demonstrated that arguments based on nodal patterns permit the rationalisation of the relative energies of four key frontier π-MOs, and that the analysis of MCD signs and $\mathcal{B}_0/\mathcal{D}_0$ ratios can play a key role in validating theoretical descriptions of the electronic structures of low symmetry porphyrinoids.

10.3.5 Analysis of the MCD Spectra of Benzofuran Derivatives Substituted with Group 16 Heteroatoms[19]

The physicochemical nature of a group of aromatic heterocycles derived from the indenide anion was studied to obtain a clear-cut picture of how the incorporation of a group 16 heteroatom affects their spectroscopic properties. The excitation energy for the lowest energy bands in the spectra of benzofuran derivatives shifts to the red, in proportion to the size of the heteroatom. Analysis of CD and MCD spectra, coupled with TD-DFT calculations, enabled definitive band assignments for these heterocyclic systems, Figure 10.14. The two L bands were identified unambiguously, using MCD spectroscopy, while the B bands were assigned using the ICD signals of their

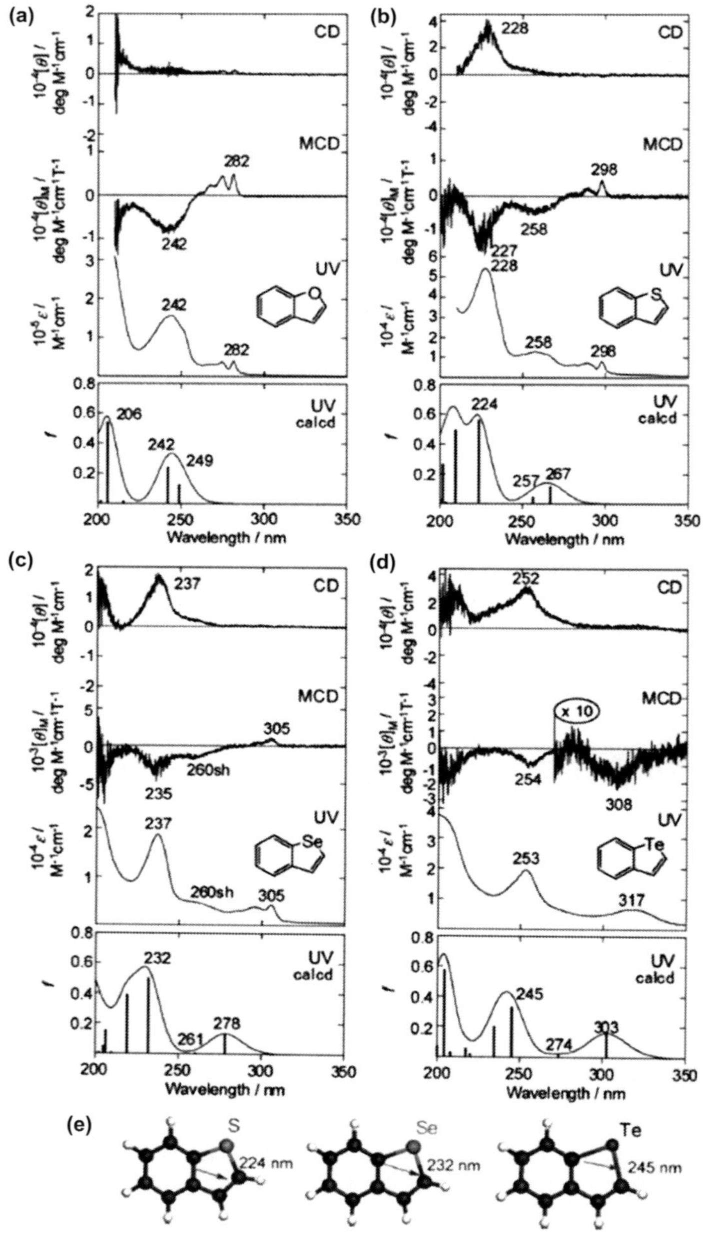

Figure 10.14 Experimental CD (top), MCD (centre) and electronic absorption (bottom) spectra of the β-cyclodextrin complexes of benzofuran (a), benzothiophene (b), benzoselenophene (c) and benzotellurophene (d) in aqueous solution at room temperature. Calculated electronic absorption spectra of the benzofuran derivatives obtained using the TD-DFT method (B3LYP/6–31 + G* (3–21G** for the heteroatom)) are shown below the experimental data. Gaussian-shaped bands with a half-band-width of 3000 cm^{-1} were used to simulate the spectra. (e) Direction of the electric dipole moment for the B_1 band of each compound.

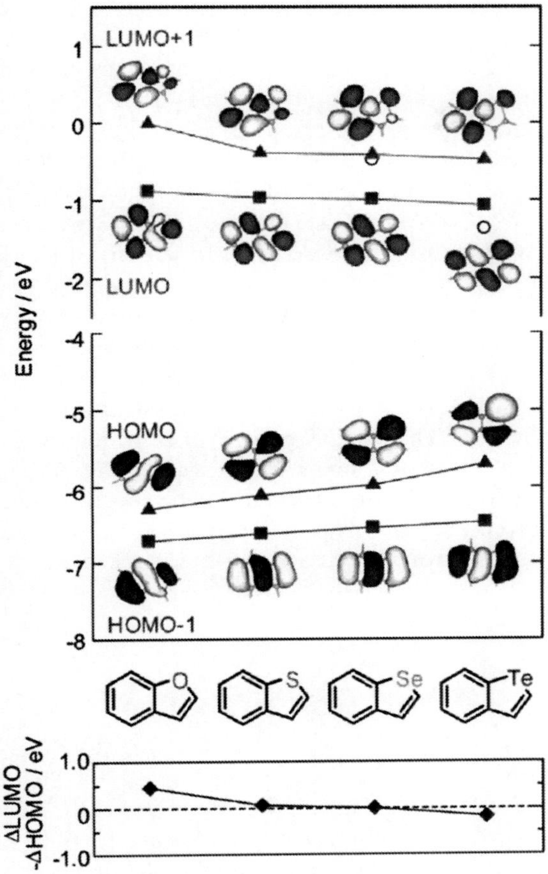

Figure 10.15 (a) Energy levels of four key frontier π-MOs of benzofuran, benzothiophene, benzoselenophene and benzotellurophene and their isosurface plots at 0.02 a.u. The white circles indicate energy levels, which are not derived from the indenide anion π-system. (b) Energy difference between the ΔHOMO and ΔLUMO values. The values were obtained from DFT calculations carried out at the B3LYP/6–31 + G* (3–21G** for the heteroatom) level of theory.

β-cyclodextrin complexes. In 1980, Michl and coworkers[20] predicted that ΔHOMO < ΔLUMO for the series of heterocyclic molecules that are obtained by replacing the CH^- of an indenide anion by N^-, NH, NR, O, S, Se and any other heteroatom which is less electronegative than CH^-. Figure 10.15 provides the ΔHOMO and ΔLUMO values that are predicted for benzofuran and its analogues with S, Se and Te heteroatoms. The +, − sign sequence with ascending energy observed for the two L bands in the MCD spectrum of benzofuran is consistent with Michl's prediction. The ΔLUMO − ΔHOMO value was found to decrease with increasing heteroatom size, owing to a destabilisation of the HOMO, and in the context of benzotellurophene the calculations predicted that ΔHOMO > ΔLUMO. A −, + sign sequence would be anticipated for the two L transitions on this basis, as observed in Figure 10.14. The intensity of the L_2 band decreases as the size of the heteroatom increases, as would be anticipated on the basis of Michl's perimeter model, as a result of the decrease of the ΔHOMO value to approach that of the ΔLUMO value, Figure 10.15, because the L_2 band would be expected to be fully forbidden when ΔHOMO = ΔLUMO. The group 16 derivatives can, therefore, be viewed as soft MCD chromophores, since the observed sign sequence of the Faraday $\mathcal{B}_0$ terms associated with the L(0,0) bands can change as a result of the structural perturbation introduced by the heteroatom. Michl and coworkers had predicted that the relative magnitudes of the ΔHOMO and ΔLUMO values of indole compounds could potentially be reversed by a particularly strong structural perturbation and that the observed MCD sign sequence could be reversed on this basis.[20] The MCD spectrum of benzotellurophene provides the first practical example of this for an indole-type molecule, and the close correspondence between the trends observed in the spectral data and in the theoretical calculations in this regard helped to validate the DFT based descriptions of the electronic structures of these molecules.

References

1. J. R. Platt, *J. Chem. Phys.*, 1949, **17**, 484.
2. (a) W. Moffitt, *J. Chem. Phys.*, 1954, **22**, 320; (b) W. Moffitt, *J. Chem. Phys.*, 1954, **22**, 1820.
3. (a) J. Michl, *J. Am. Chem. Soc.*, 1978, **100**, 6801; (b) J. Michl, *Tetrahedron*, 1984, **40**, 3845; (c) J. Michl, *Pure Appl. Chem.*, 1980, **52**, 1549.
4. J. Mack, M. J. Stillman and N. Kobayashi, *Coord. Chem. Rev.*, 2007, **251**, 429.
5. U. Höweler, J. W. Downing, J. Fleischhauer and J. Michl, *J. Chem. Soc., Perkin Trans.*, 1998, **2**, 1101.
6. J. Fleischhauer, U. Höweler and J. Michl, *Spectrochim. Acta A*, 1999, **55**, 585.
7. J. Fleischhauer, U. Höweler and J. Michl, *J. Phys. Chem. A*, 2000, **104**, 7762.
8. J. Fleischhauer and J. Michl, *J. Phys. Chem. A*, 2000, **104**, 7776.
9. J. Fleischhauer, U. Höweler, J. Spanget-Larsen, G. Raabe and J. Michl, *J. Phys. Chem. A*, 2004, **108**, 3225.
10. J. Fleischhauer, G. Raabe, K. A. Klingensmith, U. Höweler, R. K. Chatterjee, K. Hafner, E. Vogel and J. Michl, *Int. J. Quantum Chem.*, 2005, **102**, 925.
11. K. Ishii, N. Kobayashi, T. Matsuo, M. Tanaka and A. Sekiguchi, *J. Am. Chem. Soc.*, 2001, **123**, 5356.
12. R. E. Haddad, S. Gazeau, J. Pécaut, J.-C. Marchon, C. J. Medforth and J. A. Shelnutt, *J. Am. Chem. Soc.*, 2003, **125**, 1253.

13. J. Mack, Y. Asano, N. Kobayashi and M. J. Stillman, *J. Am. Chem. Soc.*, 2005, **127**, 17697.
14. S. Mori and A. Osuka, *J. Am. Chem. Soc.*, 2005, **127**, 8030.
15. S. Mori, K. S. Kim, Z. S. Yoon, S. B. Noh, D. Kim and A. Osuka, *J. Am. Chem. Soc.*, 2007, **129**, 11344.
16. A. Muranaka, O. Matsushita, K. Yoshida, S. Mori, M. Suzuki, T. Furuyama, M. Uchiyama, A. Osuka and N. Kobayashi, *Chem. Eur. J.*, 2009, **15**, 3744.
17. J. Michl, *J. Am. Chem. Soc.*, 1978, **100**, 6812.
18. (a) J. D. Keegan, A. M. Stolzenberg, Y. C. Lu, R. E. Linder, G. Barth, A. Moscowitz, E. Bunnenberg and C. Djerassi, *J. Am. Chem. Soc.*, 1982, **104**, 4305; (b) J. D. Keegan, A. M. Stolzenberg, Y. C. Lu, R. E. Linder, G. Barth, A. Moscowitz, E. Bunnenberg and C. Djerassi, *J. Am. Chem. Soc.*, 1982, **104**, 4317; (c) C. Djerassi, Y. Lu, A. Waleh, A. Y. L. Shu, R. A. Goldbeck, L. A. Kehres, C. W. Crandell, A. G. H. Wee, A. Kniezinger, R. Gaete-Holmes, G. H. Loew, P. S. Clezy and E. Bunnenberg, *J. Am. Chem. Soc.*, 1984, **106**, 4241; (d) R. A. Goldbeck, *Acc. Chem. Res.*, 1988, **21**, 95.
19. A. Muranaka, S. Yasuike, C.-Y. Liu, J. Kurita, N. Kakusawa, T. Tsuchiya, M. Okuda, N. Kobayashi, Y. Matsumoto, K. Yoshida, D. Hashizume and M. Uchiyama, *J. Phys. Chem. A*, 2009, **113**, 464.
20. M. A. Souto, S. L. Wallace and J. Michl, *Tetrahedron*, 1980, **36**, 1521.

Subject Index

Note: References to figures are indicated by *f*. References to tables are indicated by *t*.